01-CTS-21T

Post-Project Monitoring of BMPs/SUDS to Determine Performance and Whole-Life Costs

by:

Les Lampe
Howard (Andy) Andrews
Black & Veatch Corporation

Michael Barrett
Center for Research in
Water Resources
University of Texas

Bridget Woods-Ballard
H.R. Wallingford, Ltd., U.K.

Peter Martin
Black & Veatch
Consulting Ltd.

Charles Glass
ETEC, U.S.

Neil Weinstein
Low Impact Development
Center, Inc., U.S.

Chris Jefferies
Urban Water Technology Center
University of Abertay, U.K.

2004

WERF

Awwa Research Foundation
Advancing the Science of Water®

UKWIR
UK Water Industry Research Limited

The Water Environment Research Foundation, a not-for-profit organization, funds and manages water quality research for its subscribers through a diverse public-private partnership between municipal utilities, corporations, academia, industry, and the federal government. WERF subscribers include municipal and regional water and wastewater utilities, industrial corporations, environmental engineering firms, and others that share a commitment to cost-effective water quality solutions. WERF is dedicated to advancing science and technology addressing water quality issues as they impact water resources, the atmosphere, the lands, and quality of life.

For more information, contact:
Water Environment Research Foundation
635 Slaters Lane, Suite 300, Alexandria, VA 22314-1177
Tel: (703) 684-2470 Fax: (703) 299-0742
www.werf.org werf@werf.org

This report was co-published by the following organization. For nonsubscriber sales information, contact:

IWA Publishing
Alliance House, 12 Caxton Street
London SW1H 0QS, United Kingdom
Tel: +44 (0) 20 7654 5500
Fax: +44 (0) 20 7654 5555
www.iwapublishing.com publications@iwap.co.uk

Library of Congress Catalog Card Number: 2003108647

IWAP ISBN: 1-84339-716-1

Black & Veatch Corporation, Black & Veatch Consulting Ltd., Center for Research in Water Resources, University of Texas, ETEC, Inc., H.R. Wallingford, Ltd., Low Impact Development Center, Urban Water Technology Center, University of Abertay

The research on which this report is based was funded, in part, by the United States Environmental Protection Agency through Cooperative Agreement No. CR-827345 with the Water Environment Research Foundation (WERF). Unless an EPA logo appears on the cover, this report is a publication of WERF, not EPA. Funds awarded under the Cooperative Agreement cited above were not used for editorial services, reproduction, printing, or distribution.

ACKNOWLEDGMENTS

Report Preparation

Principal Investigator:
Les Lampe
Black & Veatch Corporation

Project Team:
Howard O. (Andy) Andrews
Black & Veatch Corporation

Michael Barrett
Center for Research in Water Resources, University of Texas

Charles Glass
ETEC, Inc.

Chris Jefferies
Urban Water Technology Center, University of Abertay, Dundee

Peter Martin
Black & Veatch Consulting Ltd.

Neil Weinstein
Low Impact Development Center

Bridget Woods-Ballard
H.R. Wallingford Ltd.

Project Subcommittee:

Tyler Richards, *Gwinnett County, GA (PSC Chair)*
Michael Borst, *U.S. Enviromental Projection Agency*
Jim Conlin, *East of Scotland Water*
Graham Fairhurst, *Telford & Wrekin Council*
Michael Pickel, *City of Philadelphia*
Larry Roesner, *Colorado State University*
Peter Spillett, *Thames Water Utilities Ltd.*

Water Environment Research Foundation Staff:

Director of Research: Daniel M. Woltering, Ph.D.
Research Program Director: Jeff Moeller, P.E.

United Kingdom Water Industry Research:

Director: Mike Farrimond

American Water Works Association Research Foundation:

Project Manager: Linda Reekie

Other Contributors:

Allegany County, Maryland
Baltimore County, Maryland
Buzzard Bay Project
Calvert County, Maryland
City of Alexandria, Virginia
City of Austin, Texas
City of Durham, North Carolina
City of Eugene, Oregon
City of Madison, Wisconsin
City of Newport News, Virginia
City of Raleigh, North Carolina
City of Sacramento, California
City of San Diego, California
Delaware Department of Natural Resources
Denver Urban Drainage and Flood Control District
Fairfield-Suisun (California) Sewer Distict
Fresno (California) Metropolitan Flood Control District
Hartford County, Maryland
Howard County, Maryland
King County, Washington
Minnehaha Creek Watershed District (Deephaven, Minnesota)
Montgomery County, Maryland
Oregon Department of Environmental Quality
Prince George's County, Maryland
Regional Water Authority (New Haven, Connecticut)
Sacramento County, California
San Diego County, California
Town of Cary, North Carolina
Vallejo (California) Sanitation and Flood Control District
Virginia Tech Occoquan Watershed Monitoring Laboratory
CIRIA (Construction Industry Research and Information Association)
Environment Agency
Office of Water Services
Scottish Environment Protection Agency
Scottish Water
Telford & Wrekin Council
Thames Water Utilities Ltd.
WaterUK

ABSTRACT & BENEFITS

Abstract:

Over the past 20 years, best management practices (BMPs) in the United States have been instrumental in reducing the detrimental impacts to receiving water quality and the exacerbated flooding caused by urbanization and stormwater drainage. More recently, use of sustainable urban drainage systems (SUDS) have started to be used in the United Kingdom. Both SUDS and BMPs attempt to mimic drainage patterns of the natural watershed and provide a degree of treatment needed to improve the quality of the water discharged to an acceptable level.

The costs of conventional stormwater collection systems are determined primarily in terms of initial capital expenditure. Long-term maintenance costs are absorbed by stormwater authorities responsible for maintaining their infrastructure as part of their "asset base." Currently, only a few such responsibilities exist for BMPs and SUDS, which generally incorporate surface components and are often dependent on landscaping rather than on traditional construction techniques, but may require significant regular maintenance. Any potential adopting organization will require guidance on the maintenance regimens of different types of systems and how such regimens translate into long-term adoption costs.

The project is being conducted in two phases. Phase 1, the subject of this report, includes a literature review and a survey of stormwater authorities and organizations in the United States and the United Kingdom to identify the most commonly used BMPs and SUDS and to determine the availability of data on their cost and performance. Phase 1 also involves establishment of protocols for whole-life costs and performance data for BMPs and SUDS. Protocols for whole-life costs will be published separately.

During Phase 1 a limited number of BMPs and SUDS authorities or sites will be selected for further monitoring and assessment during Phase 2. For Phase 2, operation of the selected BMPs and SUDS will be monitored for a year in terms of pollutant removal and hydrologic/ hydraulic efficiency, and applicability of their design criteria and maintenance regimens.

Benefits:

- Data from the monitoring of BMPs and SUDS sites will provide technical performance assessments and may identify immediate lessons for improving design and maintenance procedures.
- Improvements in design procedures and guidance on the approach and effectiveness of maintenance will enable organizations involved in the adoption of BMPs and SUDS to better assess their maintenance commitments.
- Guidance on the selection, whole-life costing, design, and maintenance of BMPs/SUDS will improve confidence in their use and performance and lead to a more widespread and appropriate uptake of these systems.

Keywords: Stormwater, best management practices, sustainable urban drainage systems, performance, whole-life costing, operation and maintenance, qualitative monitoring

TABLE OF CONTENTS

(Appendices C, D, E, and F are available upon request. See note in the Executive Summary)

LIST OF TABLES

LIST OF FIGURES

LIST OF ACRONYMS

ASCE	American Society of Civil Engineers
ATV	German Association for Water
AwwaRF	American Water Works Association Research Foundation
BMP	Best Management Practice
Caltrans	California Department of Transportation
CBLAD	Chesapeake Bay Local Assistance Department
CFR	Code of Federal Regulations
CIRIA	Construction Industry Research Information Association
CSO	Combined Sewer Overflow
CWA	Clean Water Act
CWP	Center for Watershed Protection
DEX	Dunfermline Eastern Expansion
DOE	Department of Ecology
EC	European Commission
ESA	Endangered Species Act
EU	European Union
FWHA	Federal Highway Administration
ICUD	International Conference on Urban Drainage
IWPA	Interim Wellhead Protection Area
LID	Low-Impact development
MDE	Maryland Department of Environment
MS4	Municipal Separate Storm Sewer System
MWCOG	Metropolitan Washington Council of Governments
NATS	National Traffic Services (Scotland)
NATS	National Air Traffic Services
NPDES	National Pollutant Discharge Elimination System
NURP	Nationwide Urban Runoff Program
NVPDC	Northern Virginia Planning District Commission
ODOT	Oregon Department of Transportation
Ofwat	Office of Water Services
PAN	Planning Advice Note
PGDER	Prince Georges County Department of Environmental Resources
SEPA	Scottish Environmental Protection Agency
SNIFFER	Scotland and Northern Ireland Forum for Environmental Research
SUDS	Sustainable Urban Drainage System
SUDSWP	Sustainable Urban Drainage Scotland Working Party
SWRPC	Southeastern Wisconsin Regional Planning Commission
TMDL	Total Maximum Daily Load
U.K.	United Kingdom
U.S.	United States
U.S. EPA	United States Environmental Protection Agency

UKWIR	United Kingdom Water Industry Research
WERF	Water Environment Research Foundation
WLC	Whole-Life Costing
WMI	Watershed Management Institute
WQI	Water Quality Inlet
WSDOT	Washington State Department of Ecology
WWF	Wet Weather Flow

Executive Summary

ES.1 Introduction

Over the past 20 years, the use of best management practices (BMPs) in the United States and sustainable urban drainage systems (SUDS) in the United Kingdom have been instrumental in reducing both the detrimental impacts to receiving water quality and the exacerbated flooding caused by urbanization and stormwater drainage. Both SUDS and BMPs attempt to mimic the drainage patterns of the natural watershed and also can provide a degree of treatment needed to improve the quality of the water discharged to an acceptable level.

The costs of conventional stormwater collection systems are determined primarily in terms of initial capital expenditure. Long-term maintenance costs are absorbed by stormwater authorities that are responsible for maintaining their infrastructure as part of their "asset base." No such responsibilities exist for BMPs/SUDS, which generally incorporate surface components, are often dependent on landscaping rather than on traditional construction techniques, and require substantial regular maintenance. Any potential adopting organization will require guidance on the maintenance regimes of different types of systems and how such regimes translate into long-term adoption costs.

ES.2 Scope of Work

The project is being carried out in two phases. Phase 1, which is the subject of this report, includes a literature review and a survey of stormwater authorities and organizations in the United States and United Kingdom to identify the most commonly used BMPs/SUDS and to determine the availability of cost and performance data. Phase 1 also involves establishment of protocols for whole-life costs and performance data for BMPs/SUDS. An additional outcome of Phase 1 will be the selection of a limited number of BMPs/SUDS authorities or sites for quantitative or qualitative monitoring and further assessment during Phase 2.

There are distinctions between the use of the terms SUDS and BMPs, and it is important to establish the differing management goals of these two strategies. The SUDS concept is built upon a holistic, multipurpose approach to developing an integrated stormwater management plan, which addresses pollutant reduction and flood control, while providing wildlife habitat and aesthetic amenities. A SUDS consists of structural elements that address water pollution and flooding from urban stormwater runoff in a natural way. Because SUDS are "natural" systems, their behavior can be unpredictable; nevertheless, SUDS can also be both tolerant to a certain amount of misuse and are capable of self-regeneration. There are a number of proprietary products, such as cellular storage, being offered in the United Kingdom as SUDS, but these products are typically local attenuation devices that address only quantity issues. Therefore, if proprietary products are proposed, care needs to be taken over the effects of cumulative inflows and associated maintenance risks.

The concept of a BMP is different from the concept of a SUDS in several respects. A BMP incorporates measures (practices) to reduce the pollutants in discharges from industrial facilities as well as in urban runoff. These practices may be either structural or nonstructural. An example of a nonstructural practice would be a program to educate the public of the potential impact of dumping hazardous materials into storm drains. Structural BMPs, on the other hand, often consist of ponds, swales or other constructed features designed specifically for the removal of pollutants. This focus on quality over quantity occurs mainly because the U.S. Environmental Protection Agency (U.S. EPA) has no authority in the area of flood control despite the obvious environmental impacts of the larger volumes and higher rates of runoff caused by urbanization. The design of BMPs installed in the United States is driven by performance standards to achieve certain levels of pollutant removal, and aesthetic aspects are more often than not considered of secondary importance.

This study differs fundamentally from previous studies in that it is a collaborative effort between researchers in the United States and the United Kingdom. Each group brings to the study its unique perspectives on the goals, aesthetic considerations, and financing of stormwater management facilities. This joint project fosters the synthesis of ideas and provides the synergy that was missing from previous efforts, and that are essential to development of the data critically needed for understanding the long-term implications of widespread adoption of BMPs and SUDS.

ES.3 Literature Review

The literature review identified 243 publications that report the results of studies dealing with the following types of BMPs/SUDS:

- Dry extended detention basins;
- Retention and wet ponds;
- Infiltration basins;
- Infiltration trenches and soakaways;
- Retention/irrigation systems;
- Vegetated swales;
- Vegetated buffer strips;
- Sand and organic filters;
- Water quality inlets and oil/water separators;
- Bioretention areas; and
- Porous and permeable pavements.

The literature review also revealed that despite the extensive research and studies conducted over the last 25 years, significant gaps remain in our knowledge about the performance and whole life costs of BMPs/SUDS. There is great variability of performance data, with limited common framework. Phase 2 of this study will be designed to fill those gaps by working with the owners and the agencies responsible for implementation and maintenance of these facilities. The literature review is summarized in chapter 2.0 and presented in detail in Appendix A.

ES.4 Current Status of BMPs and SUDS

Chapter 3.0 describes the current status of the BMPs in the United States and of the SUDS in the United Kingdom. In the United States, the widespread use of BMPs is primarily the result of various federal, state, and local regulatory programs, driven by two components of the Clean Water Act (1972), which was amended in 1987 to include stormwater discharges. The Act requires that states assess the condition of surface waters within their jurisdiction to determine whether they are "fishable and swimmable." Where water quality is not sufficient to sustain these beneficial uses, a total maximum daily load (TMDL) must be developed for the constituents that are contributing to the impairment of water quality. Another requirement of the Clean Water Act is that municipalities obtain National Pollutant Discharge Elimination System (NPDES) permits for stormwater discharges and develop stormwater management plans, including BMPs, for controlling the pollutants in the stormwater runoff. Since implementation of BMPs is driven almost exclusively by elements of the Clean Water Act (CWA), reduction of the pollutant loads is the near exclusive objective of their implementation, with little regard for either quantity control or amenity/habitat value, the key features of SUDS. It should be noted, however, that a number of stormwater detention basins have been built because of local and state needs for hydraulic controls before the CWA.

The pending implementation of the Water Framework Directive in Europe has caused a shift in thinking toward more ecological objectives in planning river systems. Member nations of the EU must have major initiatives in place to control pollution by 2004, in particular, the watercourse degradation caused by diffuse pollution must be addressed. In the United Kingdom, SUDS are seen as a key element of such initiatives.

They are a relatively new development in urban drainage and the concepts and technology are still developing. Progress in Scotland has been more rapid than in England and Wales, and all plans for new developments in Scotland must now incorporate SUDS in their drainage strategies and designs. A great deal of public education and technical training has been provided and a considerable amount of research has been commissioned to investigate system performance, maintenance, and associated costs. In Scotland the use of SUDS is widespread and is increasing rapidly as a result of primarily the policies of the Scottish Environment Protection Agency (SEPA) and its predecessor organization, the Forth River Purification Board. At the end of 2001 there were 3913 SUDS components on 767 sites around Scotland.

In England and Wales, the environmental regulator (the Environment Agency) is responsible for defense against flooding, which accounts for a high proportion of its spending. Although the agency has powers to address the water quality issues arising from diffuse pollution, until recently this has not been high priority, and SUDS have been implemented only in some local areas. The water cycle in England and Wales is managed by private utility companies working within a highly regulated regime. Economic and customer regulation come from the Office of Water Services (Ofwat), and environment regulation from the Environment Agency, which is actively promoting the implementation of SUDS in all new developments, and a SUDS working group has published a document, *Framework for Sustainable Drainage*

Systems (SUDS) in England and Wales. This is undergoing a period of widespread consultation which concluded on August 15, 2003.

ES.5 Whole-Life Costing

Whole-life costing (WLC) of BMPs/SUDS is a method of identifying future costs and referring them back to present-day values using standard accounting techniques. WLC is recognized as an appropriate technique for valuing the total costs of assets that have regular operating costs and/or recurrent maintenance costs, based on formalized maintenance programs.

Reported construction costs vary widely, and little is known about the factors that contribute to higher costs. In addition, although long-term maintenance costs can be an important component of the whole life cost of the facility, maintenance activities of BMPs/SUDS have been rarely adequately recorded and substantiated. To address this deficiency, the WLC Protocol was being developed in Phase 1 and will be provided as a separate document.

ES.6 BMPs/SUDS Performance

Although many studies have been conducted to assess the ability of stormwater BMPs/SUDS to reduce pollutant concentrations and loadings in stormwater, and some of these studies have done an excellent job of describing the effectiveness of specific devices, there is a lack of protocols for consistent assessment of the performance of BMPs/SUDS. Although quantitative monitoring may provide evidence of performance, it cannot address issues, such as amenity objectives, which are in important aspect of SUDS and prohibitive costs often rule out its use. An important objective of this study was to develop qualitative monitoring protocols that would allow assessment of whether a BMP or SUDS is performing as designed. These protocols, described in detail in Chapter 4.0, use structured empirical observations to determine whether structural- or maintenance-related failure has occurred. This will allow managers of stormwater programs to determine, relatively inexpensively, the current status of their stormwater facilities.

ES.7 Survey of Existing Information

In the United States, more than 80 flood control districts and environmental regulatory agencies were surveyed to gain information on the use of BMPs. The list of organizations and contact persons in the appendices was compiled based on knowledge of the project team, specialty conference mailing lists, and other sources. Of the approximately 50 organizations that responded, 31 appeared to have data applicable to this study. All but one of the 31 organizations were using two or more of the BMPs of interest to this study. Twelve organizations indicated that they had capital and operation and maintenance cost data; 15 stated that they were conducting routine monitoring of their BMPs, and 14 organizations indicated that they were conducting assessments of conditions downstream from their BMPs.

It is anticipated that approximately 20–30 sites in England and approximately 30 sites in Scotland will be selected for qualitative monitoring during Phase 2. Sites available for selection are listed in an appendix and summarized in the table below. The spread of sites reflects the frequency of use of different SUDS.

Table ES-1. Potential Sites for Qualitative Monitoring.

	Sites in England	Sites in Scotland
Swales	10	12
Permeable pavements	11	4+
Filter drains	4	10
Filter strips	1	0
Soakaways	7	3
Retention ponds/wetlands	15	7
Detention basins	4	11
Total	52	47

Final selection of the participating U.S. and U.K. organizations and monitoring sites will be made at the beginning of the Phase 2 study. Chapter 5.0 describes in detail site selection criteria and the candidate sites for monitoring during Phase 2.

Please Note: Appendices C, D, E, and F of this report are available through WERF and/or the third-party publisher listed on page ii and the back cover of this report. For more information, call the organization from which you purchased this report or contact WERF at 703-684-2470 or visit our website at www.werf.org.

CHAPTER 1.0

INTRODUCTION

1.1 Background

Over the past 20 years, best management practices (BMPs) in the United States have been instrumental in reducing the detrimental impacts to receiving water quality and the exacerbated flooding caused by urbanization and stormwater drainage. More recently, use of sustainable urban drainage systems (SUDS) has been initiated in the United Kingdom. Both SUDS and BMPs attempt to mimic drainage patterns of the natural watershed and provide a degree of treatment needed to improve the quality of the water discharged to an acceptable level.

The costs of conventional stormwater collection systems are determined primarily in terms of initial capital expenditure. Long-term maintenance costs are absorbed by stormwater authorities responsible for maintaining their infrastructure as part of their "asset base." Currently, only a few such responsibilities exist for BMPs and SUDS, which generally incorporate surface components and are often dependent on landscaping rather than on traditional construction techniques, but may require significant regular maintenance. Any potential adopting organization will require guidance on the maintenance regimens of different types of systems and how such regimens translate into long-term adoption costs.

The project is being conducted in two phases. Phase 1, the subject of this report, includes a literature review and a survey of stormwater authorities and organizations in the United States and the United Kingdom to identify the most commonly used BMPs and SUDS and to determine the availability of data on their cost and performance. Phase 1 also involves establishment of protocols for whole-life costs and performance data for BMPs and SUDS. Protocols for whole-life costs will be published separately.

During Phase 1 a limited number of BMPs and SUDS authorities or sites will be selected for further monitoring and assessment during Phase 2. For Phase 2, operation of the selected BMPs and SUDS will be monitored for a year in terms of pollutant removal and hydrologic/hydraulic efficiency, and applicability of their design criteria and maintenance regimens.

The impact of maintenance on the hydraulic and pollutant removal efficiencies of the BMP/SUDS will also be assessed in Phase 2 and their capital and operation and maintenance costs will be determined. The costs include renovation or replacement at the end of useful life to identify the true costs of these facilities to the community or utility.

1.2 Expected Benefits

This project is expected to produce the following significant benefits related to the use of BMPs and SUDS:

- In the short term, feedback from monitoring the demonstration sites will provide data for technical performance assessments and may identify ways to improve their design features and maintenance procedures.
- Short-term impacts of the monitoring results can also be measured in terms of awareness and perception of the project within the industry. Benefits will include improvement in design approach and guidance on the maintenance procedures and effectiveness. The monitoring results will also enable organizations contemplating the adoption of BMPs/SUDS to make a better assessment of their maintenance commitments. These benefits will be measured in terms of usefulness of the project outputs in program development.
- In the long term, the project should generate valuable guidance on the selection, costs, design, and maintenance of BMPs/SUDS and greater confidence in their effectiveness, and thus lead to more widespread use.

CHAPTER 2.0

SUMMARY OF LITERATURE REVIEW OF BMPs/SUDS

2.1 Executive Summary of Literature Review

The literature review of SUDS/BMPs identified 243 publications reporting results of studies for the 10 types of facilities identified for review and described below. These publications were reviewed to summarize the available information on siting, design, performance, and cost. Based on these reviews, data gaps and research needs for each of the SUDS/BMP technologies were identified. The complete literature review is reproduced in Appendix A of this report.

Dry extended detention basins are normally vegetated depressions that are dry except during and immediately following storm events when stormwater runoff is temporarily stored, allowing sedimentation of particles and attached pollutants.

Five U.K. and 12 U.S. publications on this technology were reviewed. These facilities are widely used in both United States and the United Kingdom. These basins have few siting requirements and little headloss, which make them adaptable to a number of settings. Most published reports document moderate removal of particulates and associated pollutants; however, removal of nutrients and other dissolved constituents is generally very low. Although maintenance guidelines focus on periodic sediment removal and clearing the outlet structure, the vast majority of maintenance consists of vegetation management, primarily grass mowing. The frequency of mowing depends more on considerations than on performance needs.

Siting criteria for extended detention basins are well established. Pollutant removal appears to be primarily a function of draining time, and, to a lesser degree, of basin geometry and depth. The dry extended detention basins have been the subject of numerous monitoring studies; consequently, little additional meaningful information is likely to be gained by devoting the limited resources of this project to quantitative monitoring. There is a larger uncertainty concerning construction costs, with various sources reporting wide-ranging costs. An additional area where information is needed is the minimum level of maintenance needed for proper functioning of the basins.

Retention ponds and wet ponds were the subject of articles in 15 U.K. and 32 U.S. publications reviewed. These facilities are also widely used in both the United States and the United Kingdom. Ponds are one of the best stormwater management practices incorporating the multipurpose SUDS approach of water quality, amenity, and habitat. Results of past monitoring programs indicate highly efficient solids removals and moderate to low removals of dissolved constituents for these ponds. Much of the difference between efficiency reported in the studies reviewed may be related to differences in permanent pool quality among the various sites. The wet weather discharge from ponds, especially during smaller events, consists primarily of displaced permanent pool water, which strongly affects the calculated removal efficiencies.

Research needs for ponds include the following:

- Siting criteria for safety and acceptance;
- Data on permanent pool sizing relative to the design water quality storm;
- Information on the role of emergent vegetation in pollutant removal;
- Data on the impact of vegetation harvesting on long-term nutrient reduction; and
- Long-term maintenance requirements.

Infiltration basins were the subject of four U.K. and 18 U.S. publications reviewed. These facilities are not widely used in the United Kingdom but are very common in some parts of the United States, such as the California central valley and Long Island, New York. They are especially attractive because of their potential for helping maintain the predevelopment hydrology by reducing the volume of stormwater runoff. Unfortunately, they have a history of high failure rates if located where soil and groundwater conditions are unfavorable.

Research needs for infiltration basins include the following:

- Siting criteria, because of concerns (unwarranted?) about groundwater quality impacts and early hydraulic failure;
- Data on ways to minimize clogging;
- Proof of performance in freezing weather; and
- Data on the quality of water below basins.

Infiltration trenches and soakaways were the subject of 14 U.K. and 15 U.S. publications reviewed for this report. These facilities are fairly common in the United Kingdom but less so in the United States. Like infiltration basins, they are desirable for maintaining the predevelopment hydrology. Unlike basins, they are installed below grade and are, therefore, unobtrusive. They also have had historically high failure rates.

Research needs identified in this review include the following:

- Siting criteria;
- More consistent design guidance;
- Data on groundwater quality;
- Information on construction costs; and
- Projected life expectancy.

Retention/irrigation systems are designed to promote infiltration of runoff through irrigation in areas where the soils are not permeable enough to allow installation of infiltration basins and trenches. There are no published reports on the performance of these systems and they are currently used only in environmentally sensitive watersheds in the Austin, Texas, area where stormwater discharges are strongly discouraged.

Because of the lack of published results of any studies and the rather limited use of the retention/irrigation systems, additional research is needed on several aspects of these BMPs, including the following:

- Sizing of irrigation area;
- Rate of runoff application;
- Effect of irrigation during wet weather on vegetation health;
- Construction costs;
- Maintenance frequency for pumps and irrigation system; and
- Cost of operation.

Vegetated swales are grass-lined channels designed for conveyance and treatment of stormwater. Eight U.K. and 30 U.S. publications on their performance were reviewed. The swales are common in both the United States and the United Kingdom; however, many of them are not engineered to achieve optimum pollutant removal. Although the siting criteria are well established, they can be difficult to implement in urban areas because the design is often in conflict with land development codes that require the use of curbs and gutters. Monitoring of these facilities for water quality is complicated by the absence of a single inflow point.

Research needs for these facilities include the following:

- Information on minimum length/hydraulic residence time, because current recommendations are all derived from a single study;
- Effect of grass types and height on removal efficiency;
- Reasons for the wide range of reported construction costs; and
- Maintenance requirements.

Vegetated buffer strips are grassy areas that treat runoff as diffuse overland flow. There is abundant literature for these facilities in the agricultural sector; however, only six U.S. publications on the use for urban runoff were identified in this review. The vegetated strips are relatively common in the United States although they are generally not engineered for optimum runoff treatment. One of the most common examples is highway shoulders. They have received relatively little recognition for their performance but are generally robust and likely to become channelized compared to other methods of constructing highway shoulders.

Research needs for buffer strips include the following:

- Information on sizing criteria relative to drainage area;
- Effect of grass types and height on removal efficiency; and
- Cost data—limited data are available because buffer strips may be considered to be a no-cost item if landscaping is required anyway.

Sand filters and organic filters were the subject of 29 U.S. publications reviewed for this project. These devices are widely used around Austin, Texas, and in some other parts of the United States but are unknown in the United Kingdom. There are a wide variety of configurations and types of media and highly efficient removals of solids and associated pollutants have been reported. Because of the relatively low rate of the passing runoff through these filters, they achieve substantial attenuation of peak runoff rate.

The siting, design, and performance of sand filters are well documented; however, there is little information on maintenance frequency and its relationship to solids loadings. Maintenance costs are another area where additional data is needed.

Water quality inlets and oil/water separators were the subject of 11 U.S. publications identified in the literature review. Water quality inlets, such as catch basins, are very common in the United States, especially in the Northeast, and gully pots are common in the United Kingdom. Most of the available data indicated that they are not very effective for water quality improvement, and no further research into these devices is recommended for this project.

Bioretention areas were the subject of 11 U.S. publications reviewed. These facilities combine many of the aspects of other infiltration (trenches) and filtration (sand filters) technologies. While they are a relatively new development, with few known installations, they are expected to achieve high levels of pollutant removal. By substantially reducing the volume and rate of stormwater runoff, they help maintain predevelopment hydrology. The priority research areas include long-term performance and routine maintenance requirements.

Porous and permeable pavements were the subject of studies reported in 16 U.K. and two U.S. publications. Many of these pavements have been installed as demonstration projects, and their routine use is not widespread. The data indicate that they provide significant attenuation of peak flows and generally high pollutant removal.

Research needs include the following:

- Structural integrity issues (including performance of saturated subgrades and performance under load);
- Water quality monitoring data;
- Better design guidelines;
- Operating life expectancy; and
- Maintenance requirements to preserve permeability.

The impact of the implementation of SUDS/BMPs on downstream ecology was the subject of 13 U.S. and two U.K. publications reviewed. Although there is extensive documentation of adverse downstream impacts of urbanization, there has been little success in demonstrating the positive effects of SUDS/BMP use. Studies dealing with this subject are extremely resource-intensive, and it would not be feasible to develop additional information in this area with the limited funds available for this study.

2.2 Introduction

This chapter discusses selected BMPs/SUDS in use in the United States and the United Kingdom. Appendix B includes more detailed discussions of BMPs/SUDS, advantages and disadvantages, siting criteria, design and construction guidelines, performance, maintenance, design life, and costs.

It is important initially to distinguish between the use of the terms SUDS and BMP and to establish the differing management goals of implementing these two strategies. The concept of a sustainable drainage system was developed by the Sustainable Urban Drainage Scotland Working Party (SUDSWP, 1999) and is built upon the idea of developing an integrated stormwater management plan that addresses pollutant reduction and flood control while providing habitat and aesthetic amenities. This holistic, multipurpose approach was illustrated by D'Arcy and Frost (2001) and is reproduced in Figure 2-1.

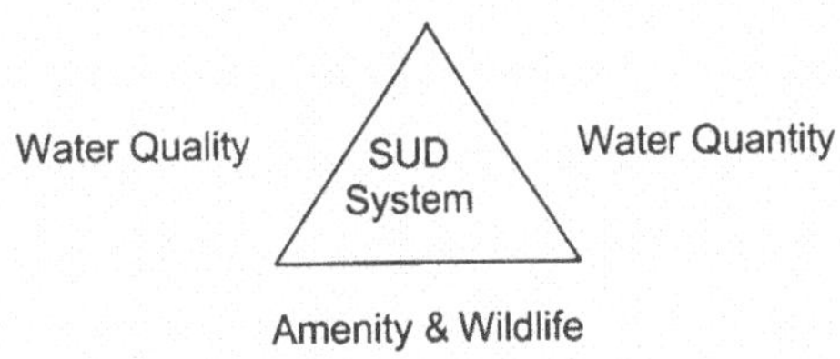

Figure 2-1. Sustainable Drainage Triangle Concept. Reprinted from *Science of the Total Environment,* Vol. 265, D'Arcy, B.J., and Frost, C.A., The Role of Best Management Practices in Alleviating Water Quality Problems Associated with Diffuse Pollution, pp 359-367 (2001), with permission from Elsevier.

As D'Arcy and Frost (2001) note, the concept recognizes that the needs of flood control engineers and pollution control scientists must be reconciled and considered together if the environment is to be adequately protected. If the resulting facilities are to be acceptable for adoption and maintenance, they must be attractively landscaped and integrated into the local community as amenity green space. Therefore, a SUDS can be defined as consisting of structural elements to address pollution and flooding from urban stormwater runoff in an attractive, natural way. The concept of sustainable drainage systems has gained popularity across Europe (Chocat, Barraud, and Alfakih, 2001).

SUDS is quite different from the best management practice, which is a central component of the U.S. Clean Water Act (CWA, 1972, and its amendments). The CWA requires the use of measures (practices) to reduce the pollutants in discharges from urban runoff and industrial facilities. These practices may be either structural or nonstructural. An example of a nonstructural practice would be a campaign to educate the public to the consequences of dumping hazardous materials into storm drains. Structural BMPs, on the other hand, consist of ponds, swales or other constructed features designed specifically for pollutant removal. Unfortunately, in some areas of the United States the definition of BMP has expanded to include facilities for flood control as well. This focus on quality over quantity occurs mainly because the U.S. Environmental Protection Agency (EPA) does not have any authority in the area of flood control despite the obvious environmental impacts of the larger runoff volumes and rates caused by urbanization. The design of BMPs installed in the United States is driven by performance standards to achieve certain levels of pollutant removal; consequently, little attention is given to aesthetics. BMPs are then sited behind buildings, fences, or other features to hide them from view.

Another significant difference between SUDS and BMPs is the configuration of existing drainage systems in urban areas. For many years, drainage systems which combined domestic and industrial wastewater with stormwater runoff were commonly used in Europe. The combined flows were discharged without treatment into receiving waters. Council Directive 91/271/EEC (May 1991) changed the ground rules by requiring secondary treatment for urban wastewater. Because about 70% of drainage systems in the United Kingdom are combined (Marsalek and Chocat, 2001), stormwater runoff from smaller events will also receive secondary treatment. Consequently, implementation of SUDS for water quality improvement is not as high a priority in the catchments served by the combined systems. Instead, there is an emphasis, particularly in England, on reducing storm runoff volumes and rates to prevent combined sewer overflows (CSOs) and any associated flooding.

Despite the difference, there is a common ground between these approaches to stormwater management, which this review and the subsequent study will address. The focus will be on stormwater management measures that are common to the United States and the United Kingdom. The following criteria were used to identify the practices included in this review:

- The practice must be a structural control, such as a wet pond, swale, or the like.
- The practice may address either water quality or quantity. Where quantity/rate control is the primary function of the practice, the emphasis will be on structures designed for smaller, more frequent events (i.e., less volume than the 2-year, 24-hour event).
- Although on-site practices may be included in the review, the focus will be on practices appropriate for 0.5-ha (1.2-ac) and larger catchments.

Using these criteria, the following structural controls were identified for review and are summarized in subsequent sections of this chapter:

- Extended detention basins;
- Wet ponds/constructed wetlands/retention basins;
- Infiltration trenches/soakaways/filter drains;
- Infiltration basins;
- Retention/irrigation;
- Sand and organic filters;
- Vegetated swales;
- Vegetated buffer strips;
- Porous pavement;
- Bioretention; and
- Water quality inlets (enhanced catch basins).

2.3 Detention Basins/Extended Detention Basins

Dry detention basins are generally vegetated depressions which are dry except during and immediately following storm events. (Wet detention basins, which have a permanent pool of water throughout the year, are discussed in the next section.) Figure 2-2 shows a detention basin used to treat highway runoff at the DEX site in Scotland.

Figure 2-2. Dry Detention Basin, DEX Site, Scotland.

During a storm, surface runoff is routed through the detention basin, and the basin outlet is blocked so that the basin fills with runoff. Detention basins are commonly used to reduce maximum runoff rates from developed areas to their predevelopment levels. The control of the maximum runoff rates serves to protect drainage channels below the basin from erosion and to reduce downstream flooding.

Dry extended detention basins are basins whose outlets have been designed to detain the stormwater runoff from a water quality design storm for a predetermined period, such as 48 hours. Extending the detention times beyond what is required for flood control improves water quality by allowing coarse silts and associated pollutants to settle (CIRIA, 2000a,b). Such systems may contain a low-flow channel and small permanent pool as a landscape feature and as a means of improving ecological benefits. A schematic of an extended detention basin with a micropool is shown in Figure 2-3.

The water quality benefits of extended detention basis include the removal of sediment and buoyant materials, as well as reduction of nutrients, heavy metals, toxic materials, and oxygen-demanding materials associated with the particles. The water quality benefits of a detention dry pond can be increased by extending the detention time. Substantial removal of total suspended solids (TSS) is possible if stormwater is retained for longer than 48 hours. Detention facilities are frequently employed for temporary sediment control during construction, and in some cases are made a permanent part of an installation (Young et al., 1996).

Dry basins can be considered where retention of large volumes of water is not beneficial or is inappropriate. It is likely that dry basins are easier to maintain than wet ponds and may be considered a lesser health and safety hazard, although the risks must be assessed for wet, dry, and filling conditions. Dry detention areas can be integrated with, for example, football pitches, making dual use of land where available open area is limited. Consideration should be given to such use in combination with other controls such as grassy swales and vegetated filter strips to achieve the required TSS removal.

2.3.1 Research Needs

On the basis of the literature review, the following research needs were identified for extended detention basins:

- Effectiveness of commonly adopted design criteria, such as detention time, length to width ratio, and optimum depth;
- Frequency of complete filling and the influence of climate change on that frequency;
- Outlet configurations that minimize maintenance;
- Integration with adjacent areas for project designers;
- Project costs, particularly clarification of the wide range of costs reported, and identification of elements that increase construction costs; and
- Empirical data related to maintenance, frequency of maintenance, and cost.

2.4 Wet Ponds/Constructed Wetlands/Retention Ponds

Wet ponds (also called stormwater ponds, retention ponds, or wet extended detention ponds) are constructed basins that contain a permanent pool of water at least throughout the wet season. They differ from constructed wetlands primarily by their greater average depth and peripheral vegetation rather than complete cover. Wet ponds and constructed wetlands treat incoming stormwater runoff through settling and biological uptake. The primary pollutant removal mechanism is settling of silts and suspended sediments. Uptake of pollutants, particularly nutrients, also occurs to some degree through biological activity in the pond. Wet ponds are among the most widely used stormwater practices. Of the several different wet pond designs, the most common is the extended detention wet pond, where storage is provided above the permanent pool in order to detain stormwater runoff and to promote settling.

Wet ponds are designed to support emergent and submerged aquatic vegetation along the shoreline, as well as an active microbial community capable of consuming dissolved constituents in the inflow. If properly designed, they also include sedimentation processes that capture the particulate fraction of the flow. Pond inflows from most runoff events replace a portion of the prior volume and are stored and treated until displaced by the perennial baseflow or next runoff event.

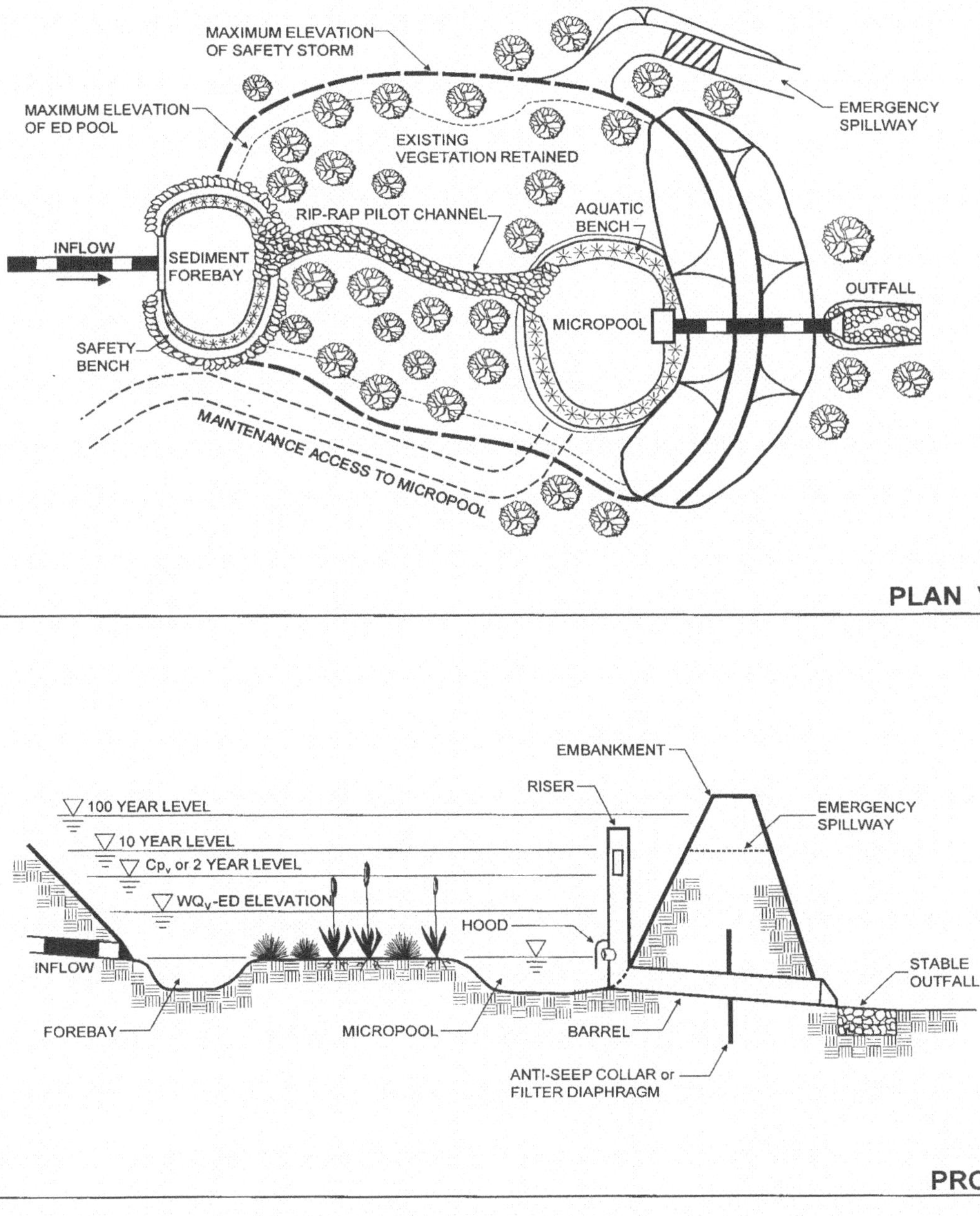

Figure 2-3. Schematic of Extended Detention Basin (Maryland Department of the Environment, 2000).

Wet pond permanent pools are generally deep enough to prevent resuspension of sediment and are sized according to a design storm runoff volume or a factor of the design treatment volume. A number of criteria are available for determining the permanent pool volume. In the U.K., the objective is to provide a retention time of 2 to 3 weeks during the wettest months. In addition to allowing time for biological treatment, this will also improve the settling and ultimate removal of suspended particulates (CIRIA, 2000a,b).

Enhanced wet ponds can be designed with a forebay from where trapped coarse sediment can be easily removed during maintenance, and with aquatic benches that support a fringe wetland along the shore. Shallow depths along the pond perimeter mitigate the potential hazard to children. Environmental benefits include creation of aquatic, wetland, and terrestrial habitats, and high community acceptance, if designed as an attractive urban amenity. A wet pond is presented in Figure 2-4, and a schematic representation of a typical wet pond design is shown in Figure 2-5.

2.4.1 Research Needs

On the basis of the literature review, the following research needs were identified for wet detention basins:

- Design guidelines for sizing the permanent pool relative to the design storm to be treated;
- Design guidelines for achieving adequate detention times across all expected storm events;
- Design guidelines for appropriate treatment volumes and detention times to protect downstream channels;
- Water quality, particularly long-term, with mass balance figures for several years. Not enough is known about the impacts that nutrient build-up in vegetation has on nutrient through-flow. The benefits of vegetation harvestings, and the best approach to harvesting, need to be addressed.
- Uptake of heavy metals;
- Pollutant removal rates during the nongrowing season.
- Rate of accumulation of sediments and increases in toxicity;
- Relationship between pond performance and appearance, and perception locally;
- Influence of the pond on residential sales and costs;
- Effectiveness of different marginal arrangements in providing safety barriers;
- Long-term maintenance costs;
- Impact of differing maintenance regimens on pond performance (hydraulics and water quality); and
- Maintenance needs associated with sediment management (which currently are largely being ignored).

Figure 2-4. Wet Pond, DEX Site, Scotland.

2.5 Infiltration Basins

Infiltration basins are vegetated stormwater retention facilities designed to capture first flush runoff and to allow it to infiltrate directly into the soil. This practice is intended to mimic the natural rainfall retention and infiltration processes. Basins may be excavated in or constructed above grade using structural walls or berms. Typical designs allow for complete infiltration of the water quality volume within 2 to 3 days and incorporate a splitter structure to route surplus inflows around the facility when it is full. Infiltration basins are generally suitable for serving drainage areas of 5 to 15 acres. Smaller areas may be better served by infiltration trenches (Schueler, 1987).

Vegetation planted in and around infiltration basins should be resistant to temporary inundation. Root penetration and thatch formation helps maintain and often enhances infiltration capacity of the basin floor. In addition, vegetation can trap suspended solids in stormwater by growing through accumulated sediments and preventing resuspension. Vegetation also provides nutrient uptake in the shallow root zone and a substructure for microbial growth. A schematic of an infiltration basin is presented in Figure 2-6.

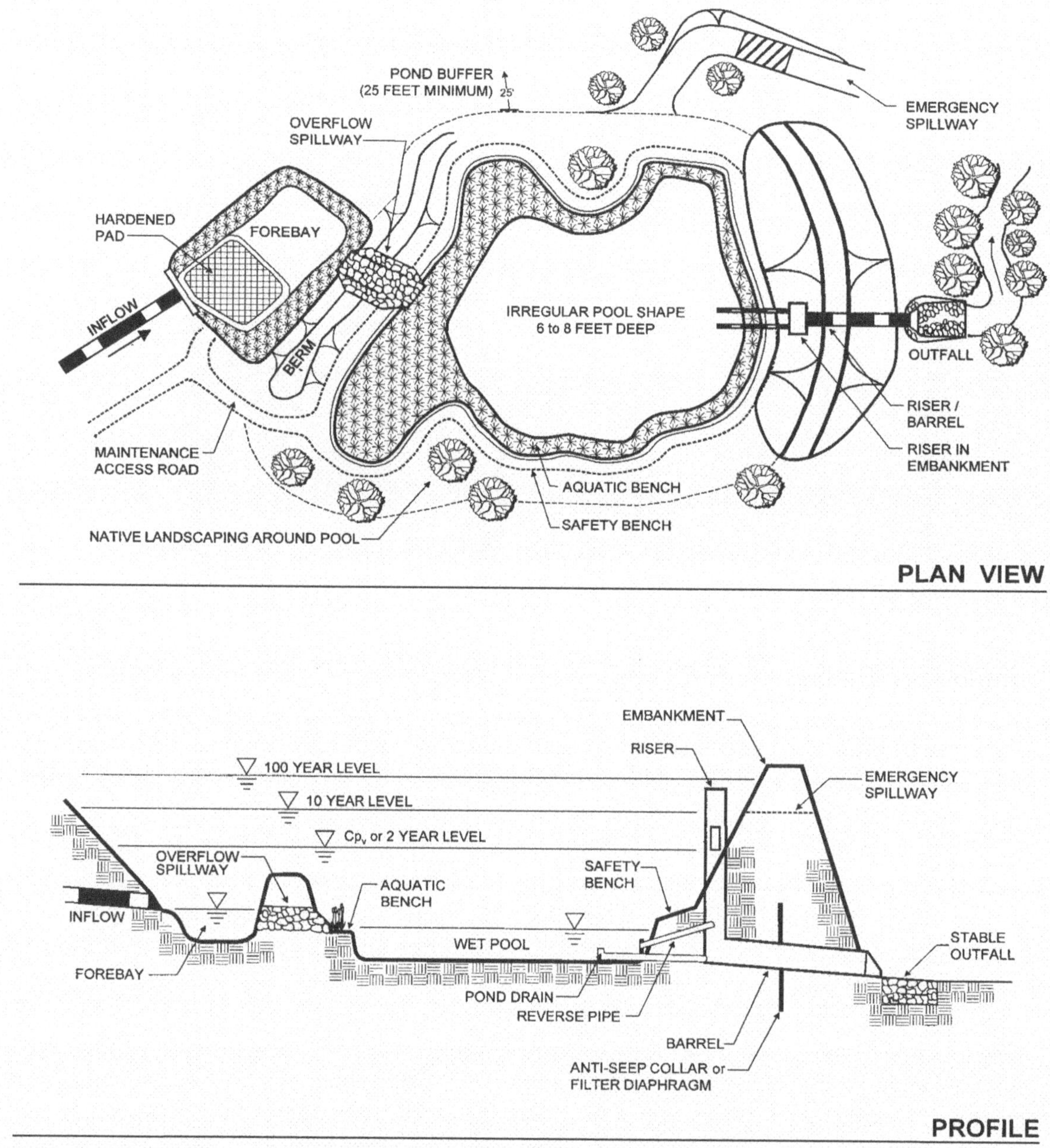

Figure 2-5. Schematic of Wet Pond (Maryland Department of the Environment, 2000).

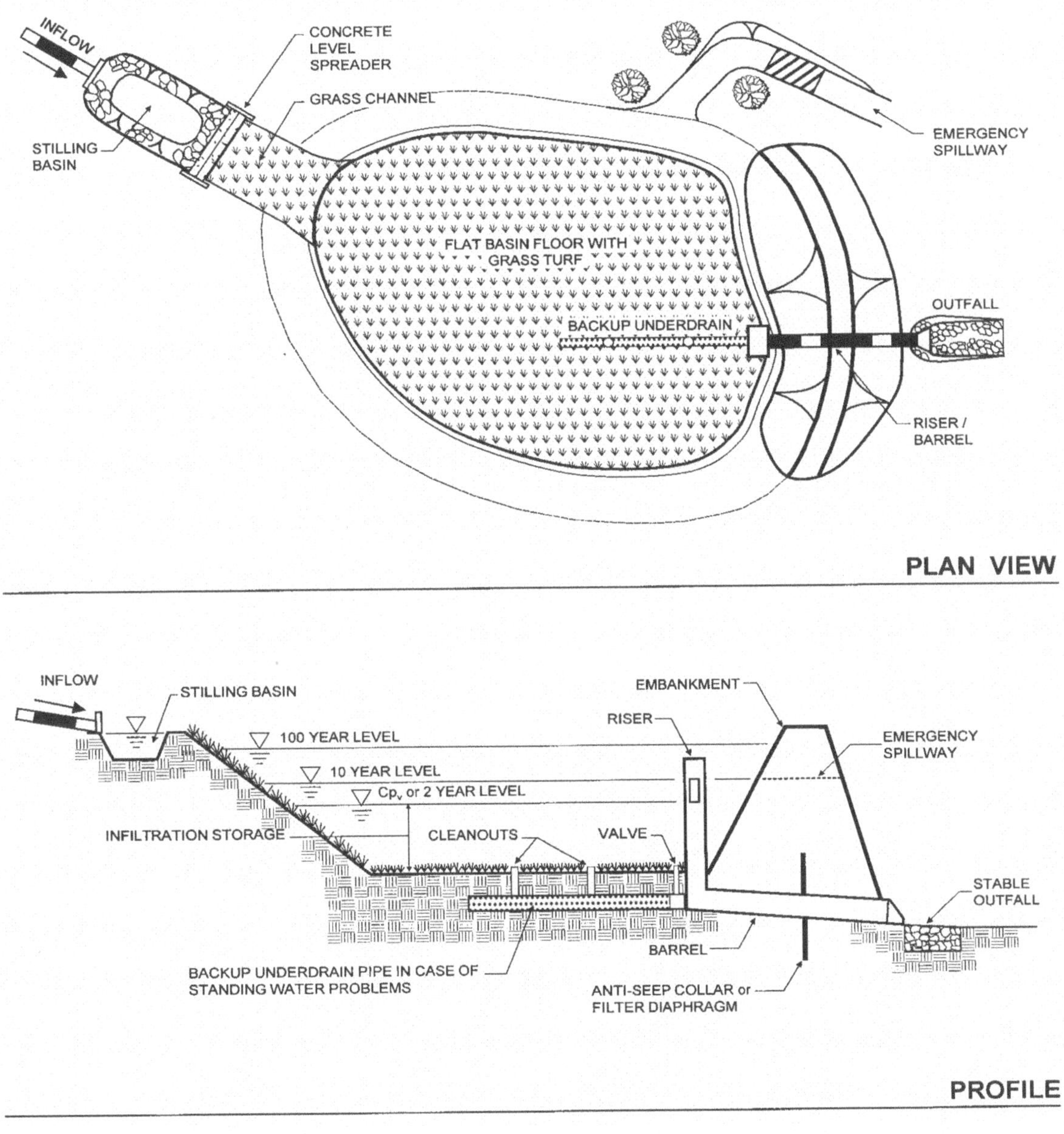

Figure 2-6. Schematic of Infiltration Basin (Maryland Department of the Environment, 2000).

2.5.1 Research Needs

The fundamental feasibility of the infiltration approach is still being questioned. The basic issue is whether the temporary, but deep, ponding in basins makes clogging inevitable. On the basis of the literature review, the following research needs for infiltration basins were identified:

- Pretreatment system that can ensure basin longevity;
- The effectiveness of basins under freezing conditions and snowmelt conditions;

- The accuracy of current field measuring methods in estimating long-term local soil infiltration rates; and
- The actual monitored pollutant removal performance of a fully operational basin, particularly with respect to soluble nutrients.

2.6 Infiltration Trenches/Soakaways/Filter Drains

Infiltration trenches are shallow excavations filled with rubble or stone to create temporary subsurface storage for infiltration of stormwater runoff. Some manufacturers have developed modular plastic systems which provide more void space than rock-filled trenches and reduce the amount of excavation required to store the water quality volume. Usually a small portion of the runoff, the first flush, is diverted to the infiltration trench. The runoff gradually exfiltrates through the bottom and/or sides of the trench into the subsoil and eventually to the water table. The primary means of pollutant removal is filtration through the soil. Figure 2-7 is a schematic of a typical end-of-pipe system. Trench designs may be modified to include vegetative cover and other features to establish a biofiltration area.

Infiltration trenches may be designed for complete exfiltration or partial exfiltration (where the first flush of the runoff is routed to the trench and the remainder is bypassed and conveyed to other BMPs).

Infiltration trenches should always be constructed to include pretreatment. Infiltration technologies should not be used in areas of potentially high pollutant loadings. In groundwater recharge areas, such as Zone II and Interim Wellhead Protection Areas (IWPA), infiltration technologies may be used for uncontaminated rooftop runoff only (Stormwater Management, 1997).

In the United Kingdom, infiltration trenches are commonly known as soakaways and can be built in different forms and sizes, such as square or circular pits, which are filled with rubble or lined with dry-joined brickwork or precast perforated concrete ring units surrounded by granular backfill; or as trenches that follow convenient contours (BRE, 1991). Soakaways are used extensively in draining small catchments in remote rural areas.

Filtration trenches or filter trenches, which include underdrain systems, combine some of the attributes of infiltration trenches and filter systems. They are used in the United Kingdom in low-permeability soils, mainly for purifying water before it is conveyed downstream. Filter trenches with impermeable liners are used in areas where the risk of groundwater contamination is high.

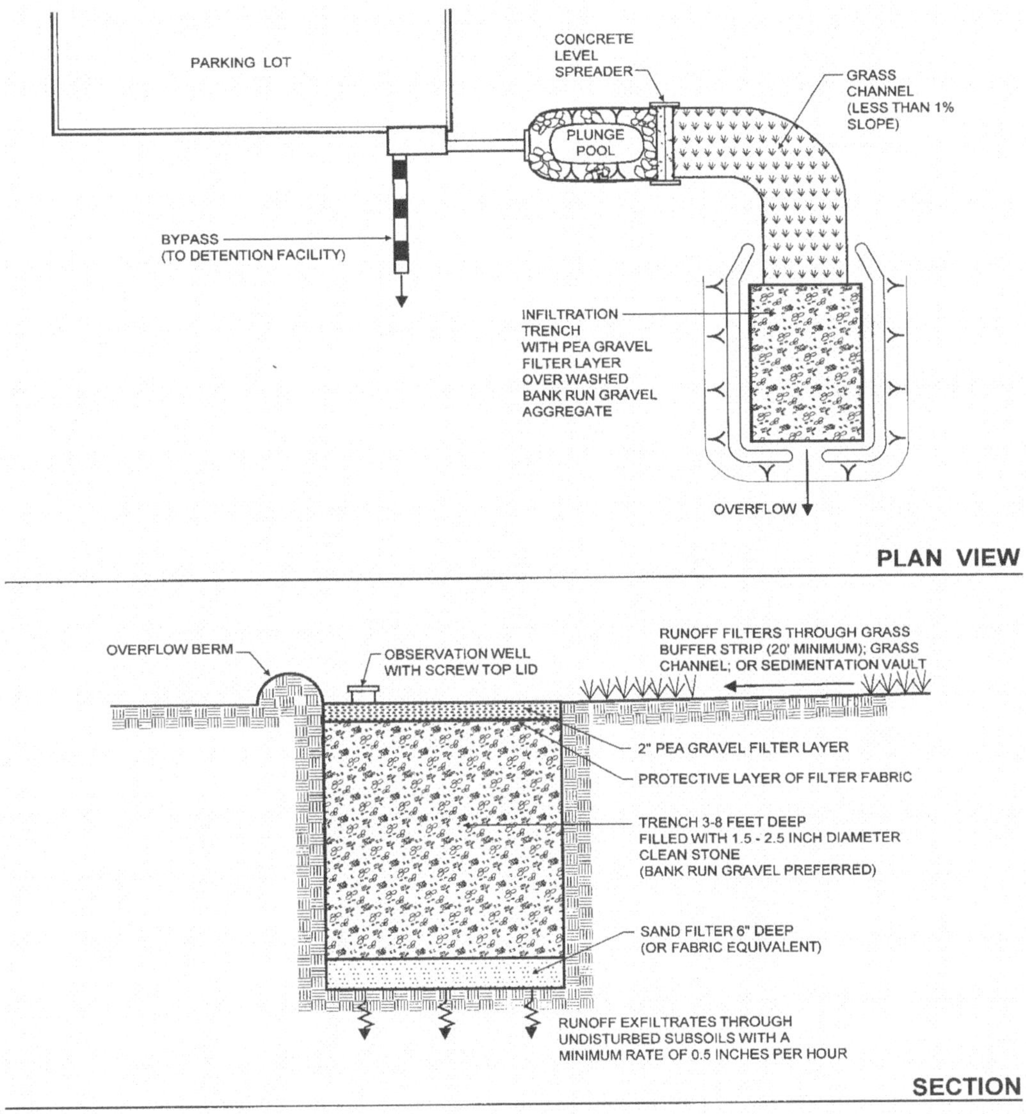

Figure 2-7. Schematic of Infiltration Trench as End-of-Pipe System (Maryland Department of the Environment, 2000).

For roadside drainage, a perforated pipe combined with an infiltration/filtration trench is known as filter drain. Typically, the filter drain is connected with a trapped gully pot, which in the United States is known as a catch basin. It is estimated that filter drains are used to collect surface water runoff from 25% of all major roads in the United Kingdom (CIRIA, 1999).

2.6.1 Research Needs

On the basis of the literature review, the following research needs were identified for wet infiltration trenches:

- Potential for groundwater contamination caused by certain land uses;
- Design guidelines for calculating the size of the trench based on permeability and surface area available for infiltration;

- ♦ Confirmation of modeling results with respect to effective drainage area (bottom and sides of trench) and effect of blockage of the bottom of the trench;
- ♦ Attenuation of contamination based on soil characteristics. Additional evaluation of the performance in sandy soils near the water table;
- ♦ Performance in subfreezing weather and during snowmelt runoff. The possibility of groundwater contamination by soluble substances in deicing salts;
- ♦ Project costs, particularly clarification of their wide range and identification of elements that drive construction costs upward; and
- ♦ Cost of trench reconstruction (versus original construction cost).

2.7 Retention/Irrigation

Retention/irrigation refers to the capture of stormwater runoff in a holding pond and its subsequent use for irrigation of pervious landscaped or natural areas. This technology is very effective because it results in virtually no discharge to receiving waters and because it has a high stormwater constituent removal efficiency. It mimics natural undeveloped watershed conditions where the majority of the rainfall from smaller events infiltrates through the soil profile. Its main advantage over other infiltration technologies is the use of an irrigation system to spread the runoff over a larger area, which makes it particularly well suited for areas of low permeability soils. A schematic of a retention/irrigation system is presented in Figure 2-8.

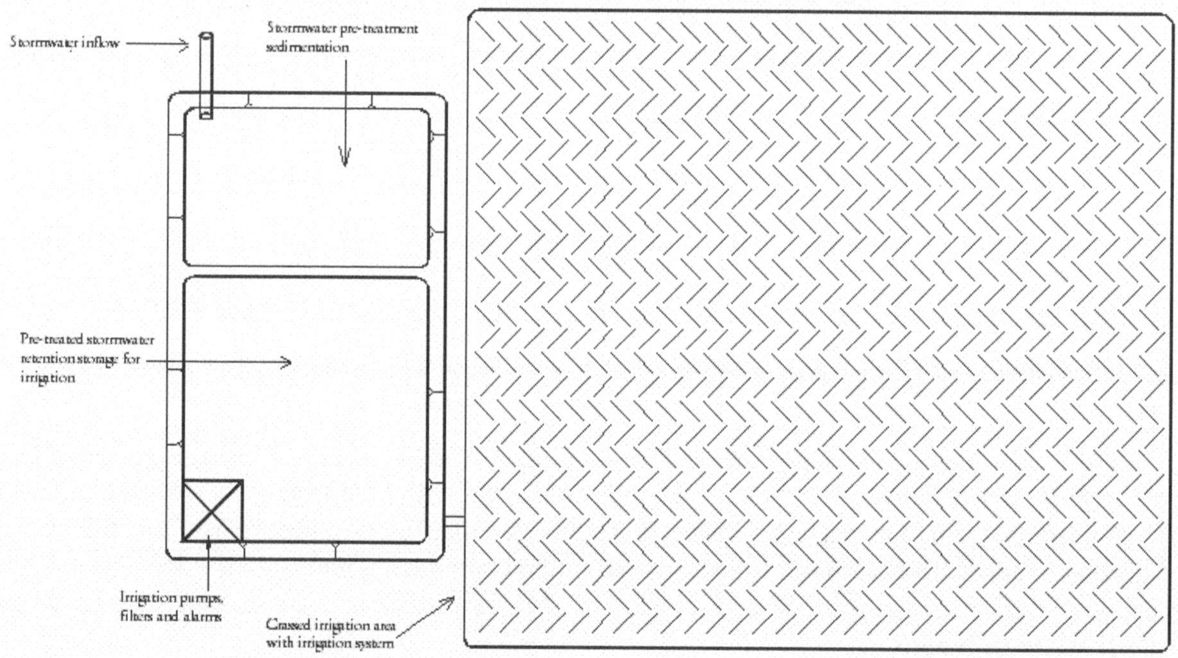

Figure 2-8. Schematic of Retention/Irrigation System.

Stormwater can be captured in almost any kind of runoff storage facility, from dry, concrete-lined ponds to vegetated basins and permanent pools. The pump and wetwell should be equipped with a rainfall sensor to provide irrigation only during periods when the required infiltration rates can be realized. Generally, a spray irrigation system is required to achieve an adequate flow rate. Collection of roof runoff for subsequent use (rainwater harvesting) is another method of retention/irrigation.

This technology is still in its infancy, and there are no published reports on its effectiveness, cost, or operating requirements.

2.7.1 Research Needs

Retention/irrigation is an emerging BMP, and research into almost all phases of design, operation, maintenance, and cost is needed. Important elements include the following:

- Sizing of the irrigation area;
- Rate of application of retained runoff;
- Construction costs;
- Maintenance frequency for pumps and irrigation system; and
- Cost of operation.

2.8 Vegetated Swales

A vegetated swale is a broad, shallow channel with a dense stand of vegetation covering the side slopes and bottom. Swales can be natural or manmade, and they are used to trap particulate pollutants (suspended solids and trace metals), to promote infiltration, and to reduce the flow velocity of stormwater runoff. A typical swale is shown in Figure 2-9 and a schematic is given in Figure 2-10. Vegetated swales can be used in place of curbs, gutters, and pipes in a stormwater conveyance system, but they are limited in the amount of runoff they can treat effectively. Therefore, swales are best suited for residential, industrial, and commercial areas with low flow and small populations.

As storm runoff flows through a vegetated swale, sedimentation is promoted by the relatively low velocity and shallow water depth. Some uptake of nutrients by the vegetation may occur and a substantial amount of the runoff infiltrates into the underlying soils. Through geometric modifications and other features, the swale can be designed to serve as both a treatment and a conveyance system.

Figure 2-9. Vegetated Swale.

In snowy climates, swales may serve the dual purpose of both snow storage and stormwater management. This is particularly relevant where swales are used to treat road runoff. If used for this purpose, they should be planted with salt-tolerant vegetation, such as creeping bentgrass.

In arid or semiarid climates, swales should be planted with drought-tolerant vegetation, such as buffalo grass. The value of vegetated practices for water quality management must be weighed against the cost of the water needed to maintain them.

2.8.1 Research Needs

On the basis of the literature review, the following research needs were identified for vegetated swales:

- Design limits and effects of varying residence times and side slopes;
- Hydrological analysis of various storm frequencies and intensities to establish the long-term viability of current designs;
- Best designs for optimal pollutant removal;
- Relationship between removal rate and age of swale;
- Effectiveness of check dams or steps in preventing downstream migration of pollutants;
- Effects of different grass types and height on pollutant removal capability;
- Project costs, particularly clarification of their wide range and identification of elements that drive construction costs upward; and
- Empirical data related to maintenance requirements, frequency, and cost.

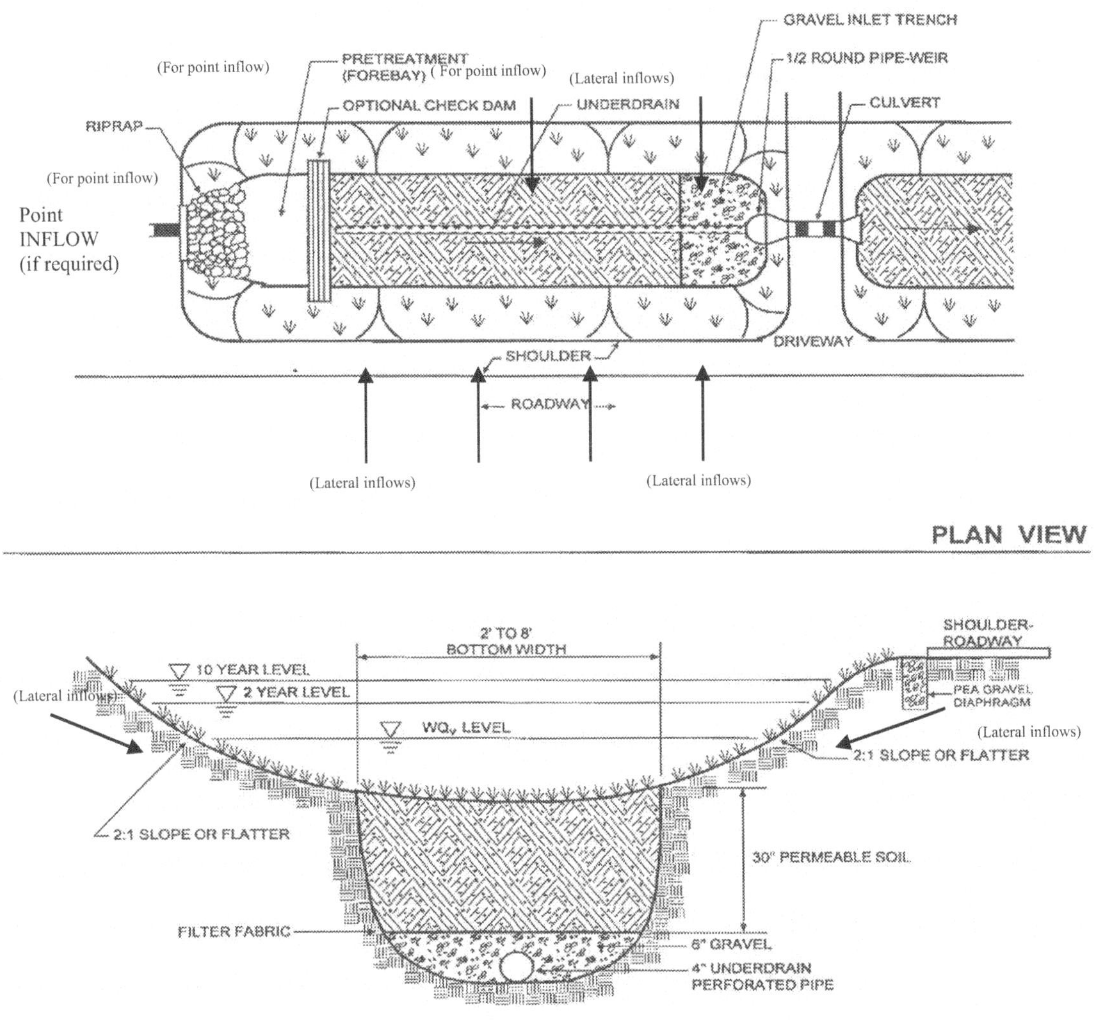

Figure 2-10. Schematic of Vegetated Swale — U.S. Example (Maryland Department of the Environment, 2000).

2.9 Vegetated Buffer Strips

Grassed buffer strips (vegetated filter strips, filter strips, and grassed filters) are designed to treat sheet flow from adjacent surfaces. Filter strips function by slowing runoff velocities and allowing sediment and other pollutants to settle, and by allowing some infiltration into underlying soils. Buffer strips were originally used as an agricultural treatment practice, and have more recently evolved into an urban practice. With proper design and maintenance, buffer strips can provide relatively high pollutant removal. Since the public views them as landscaped amenities rather than part of stormwater infrastructure, there is little resistance to their use.

Buffer strips can be divided into two basic types: natural buffers and engineered biofilters. Both types encourage diffuse, shallow, low velocity flow across a vegetated surface and both can provide stormwater quality enhancement if properly designed. Natural vegetated buffers are typically left undisturbed and are used in areas of relatively low-density development as a passive, low-maintenance means of protecting nearby receiving waters.

Engineered biofilters are designed and constructed to maximize the water quality benefits of vegetative filtration, particularly where adequate natural buffers do not exist or cannot be preserved. Engineered systems feature mechanically graded surfaces, with vegetation selected to suit physical site conditions, water quality requirements, and aesthetic appeal. Figure 2-11 presents a schematic of an engineered biofilter, and Figure 2-12 shows an installation.

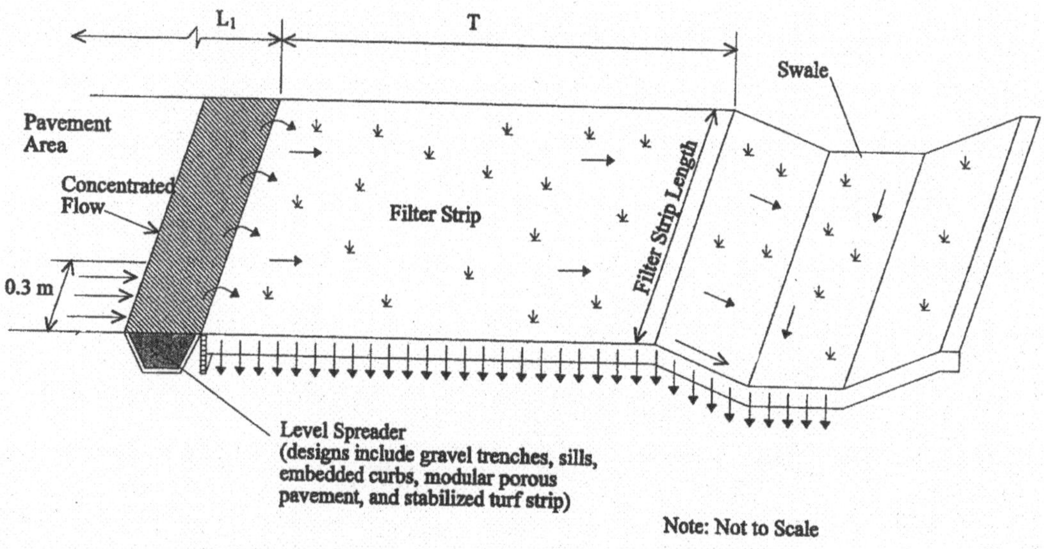

Figure 2-11. Engineered Filter Strip Schematic.

Figure 2-12. Filter Strip, Hopwood Services, U.K.

2.9.1 Research Needs

On the basis of the literature review, the following research needs were identified for filter strips:

- Size of drainage area;
- Effects of mowing on pollutant removal capability of grassed filter strips;
- Pollutant removal capability of various types of grass;
- Whether removal rates decline with age;
- Optimum mowing frequency to retain optimum grass length; and
- Construction out-turn (as-built) costs.

2.10 Sand and Organic Filters

Sand filters usually consist of two chambers: a settling chamber and a filter bed filled with sand or another filtering material. Large particles settle out in the first chamber and the finer particles and other pollutants are removed through the filtering medium.

The use of slow-sand filters for treatment of stormwater runoff is a fairly recent development. The filtration of water through sand as a means of improving its quality was first practiced in 1829 in London to treat Thames River water. The slow-sand filter, later developed in England, was used extensively at the beginning of the 20th century in the United States in both water and wastewater treatment. Over the years, the use of slow-sand filters declined as rapid filtration and other treatment technologies were developed (Young et al., 1996).

Since their original inception in Austin, Texas, intermittent sand filters have been widely used to treat stormwater runoff throughout the United States. Major types of these filters include the Austin sand filter (Figure 2-), the District of Columbia underground sand filter, the Alexandria dry vault sand filter, the Delaware sand filter (Figure 2-13), and the multichamber treatment train (Figure 2-15). The primary differences among these designs are location (i.e., above or below ground), the drainage area served, filter surface areas, land requirements, and the quantity of runoff treated. Modifications to the traditional surface sand filter were made primarily to fit them into more challenging sites (e.g., underground and perimeter filters) or to improve pollutant removal (e.g., organic media filter).

The City of Austin recognizes two filter design alternatives: "full" and "partial" sedimentation. The full sedimentation filter consists of a sedimentation basin sized to contain the entire water quality volume, which is then released over 24 hours to the filter basin. The partial sedimentation filter consists of a single basin that is underlain by the filter bed. Because the degree of pretreatment provided by the partial sedimentation filter is low, the city specifies lower permeability for the sand, and thus a larger filter area is required as a safety factor.

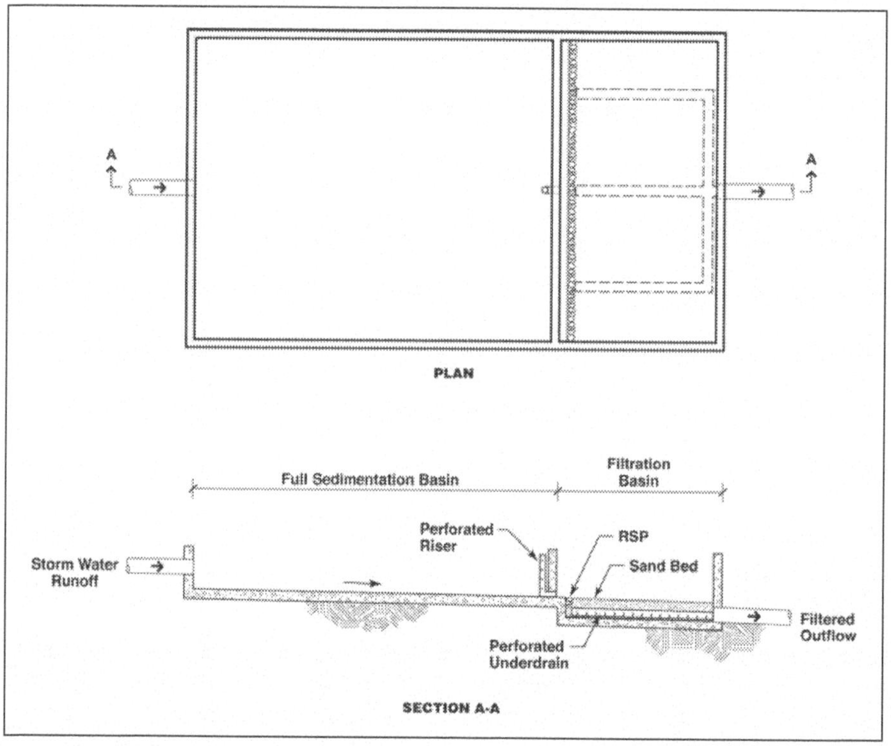

Figure 2-13. Schematic of "Full Sedimentation" Austin Sand Filter.

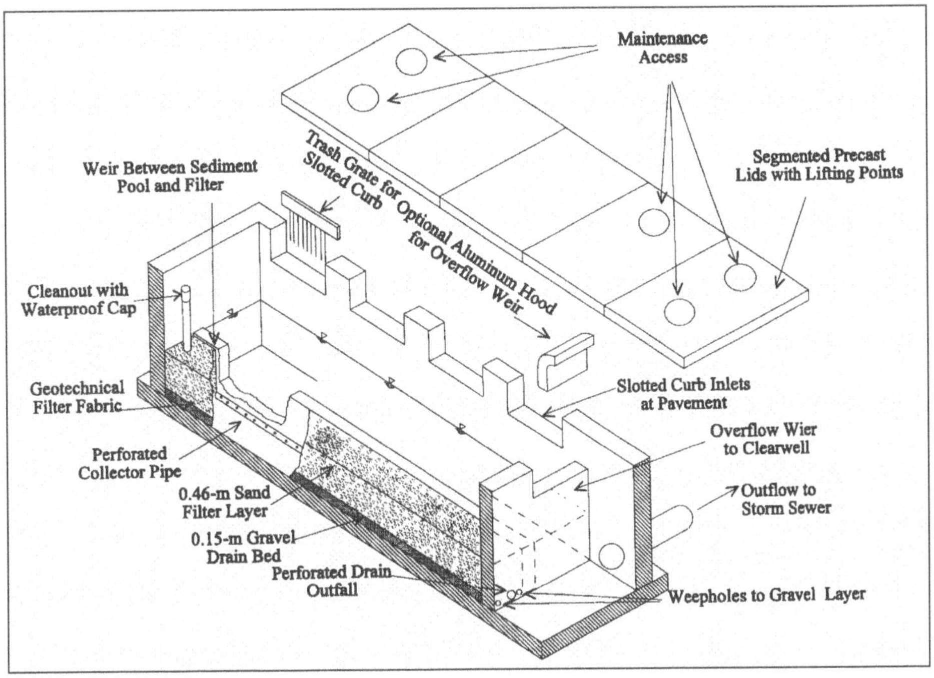

Figure 2-14. Schematic of Delaware Sand Filter (Young et al., 1996).

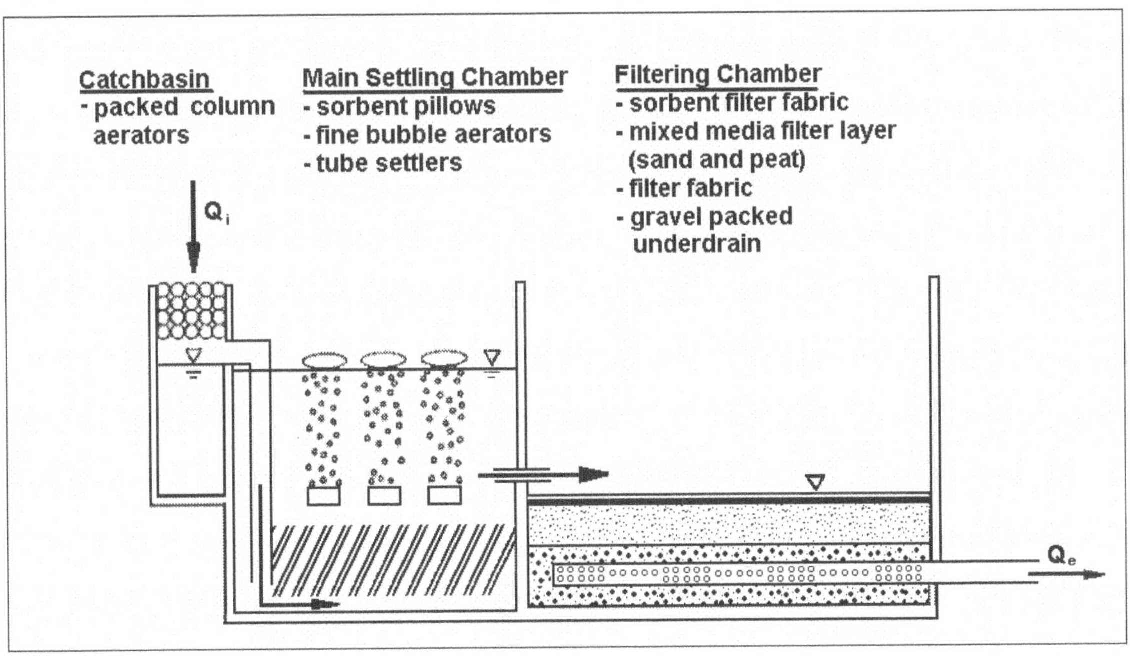

Figure 2-15. Schematic of Multichambered Treatment Train.

2.10.1 Research Needs

Sand filters have been used and monitored frequently in a few areas of the United States, so their siting criteria, design guidelines, and performance for most common constituents are well understood. However, on the basis of the literature review, the following research needs were identified:

- The use of media other than sand to improve the removal of dissolved constituents such as nutrients and metals;
- Maintenance requirements and their costs; and
- Project costs, particularly clarification of their wide range and identification of elements that drive construction costs upward.

2.11 Water Quality Inlets and Oil/Water Separators

Water quality inlets (WQIs), also called trapping catch basins, oil/grit separators, or oil/water separators, consist of one or more chambers that promote sedimentation of coarse materials and separation of free oil (as opposed to emulsified or dissolved oil) from stormwater. Some WQIs also contain screens to help retain larger or floating debris, and many of the newer designs include a coalescing unit that helps promote oil/water separation. A typical WQI, as shown in Figure 2-16, consists of a sedimentation chamber, an oil separation chamber, and a discharge chamber. In the United Kingdom, simple versions of a WQI, termed gully pots, are often incorporated into urban and highway drainage systems, but because of their small volume they are unlikely to achieve substantial removal of pollutants.

Water quality inlets and oil/water separators are appropriate for capturing hydrocarbon spills, but provide very marginal sediment removal and are not very effective for treatment of stormwater runoff. Most available data indicated that these devices are not very effective for water quality improvements, and no further research is recommended for this project.

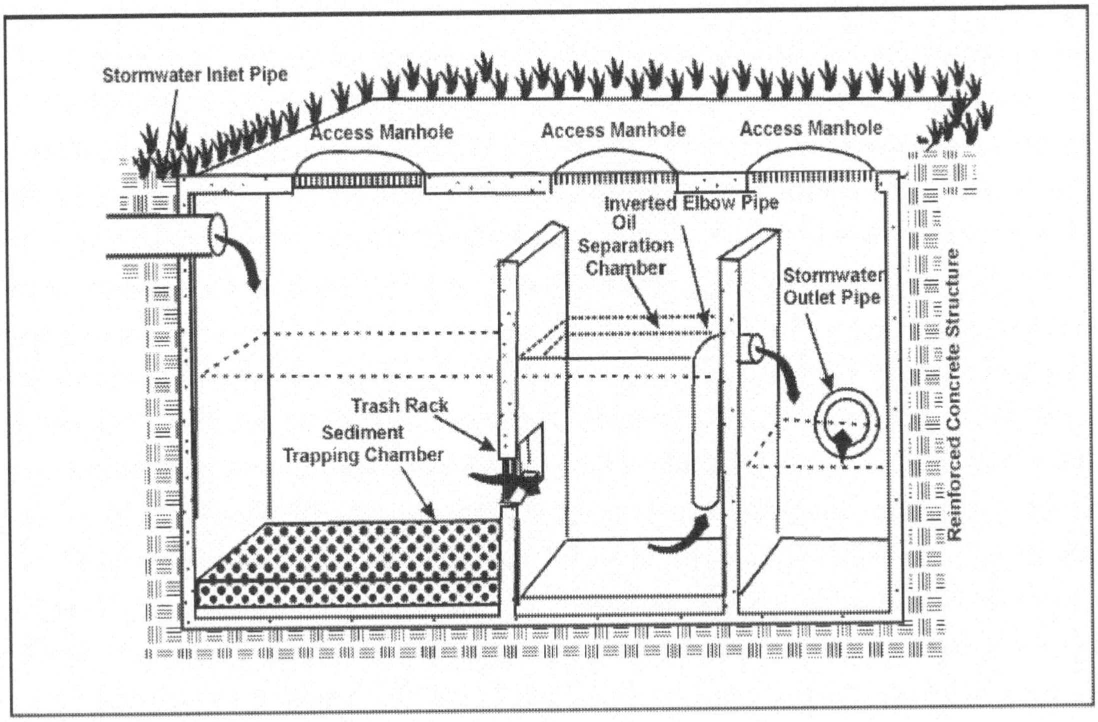

Figure 2-16. Schematic of Water Quality Inlet (Berg, 1991).

2.12 Bioretention Areas

Bioretention areas function as soil and plant-based filtration devices that remove pollutants through a variety of physical, biological, and chemical processes. As shown in the *Stormwater Technology Fact Sheet: Bioretention* (EPA, 1999d), Figure 2-17 displays runoff conveyed as sheet flow to the treatment area. A bioretention area consists of a grass buffer strip, a sand bed, a ponding area, an organic or mulch layer, planting soil, and plants. The runoff velocity is reduced by passing it over or through a sand bed and subsequently distributing it evenly over a ponding area. The ponding area, an organic surface layer and/or ground cover and the underlying planting soil, is graded and its center depressed. As the water gradually infiltrates the bioretention area or is evapotranspired, it is ponded to a depth of 6 in. (15 cm). The bioretention area is graded to divert runoff in excess of the design storm volume. Exfiltration of the water stored in the bioretention area into the underlying soils occurs over a period of days.

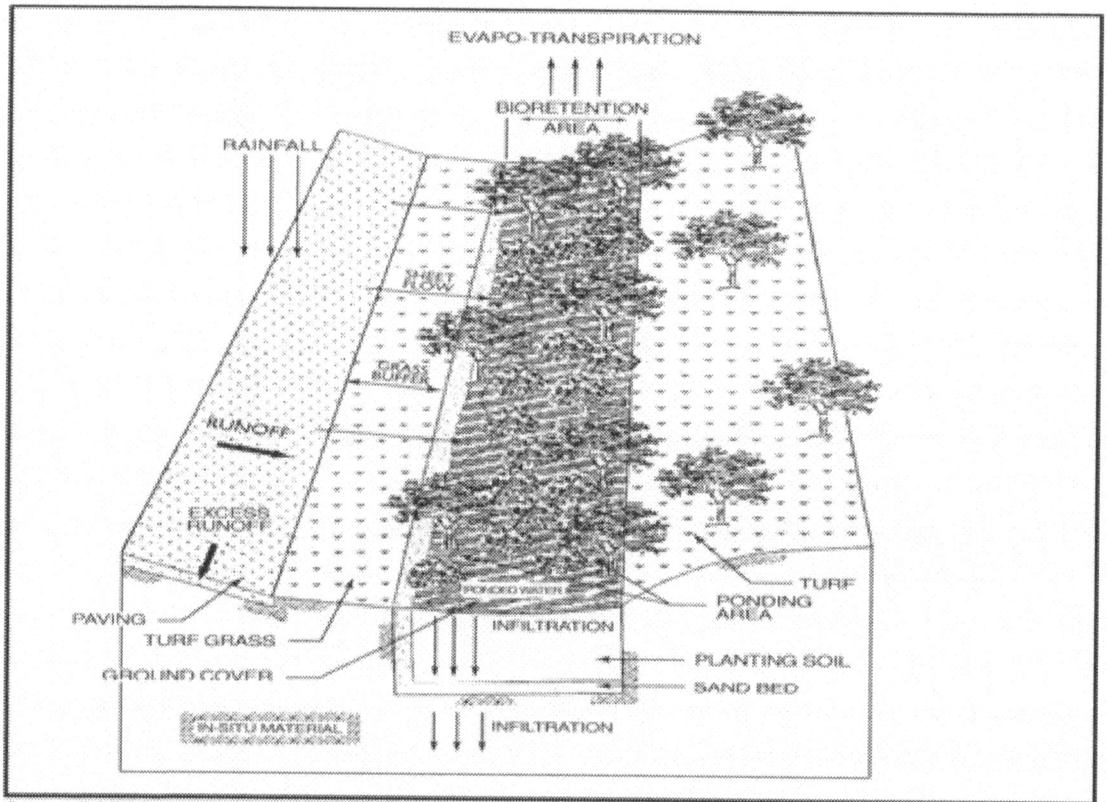

Source: PGDER, 1993.

Figure 2-17. Bioretention Area (EPA, 1999d).

Each component of the bioretention area is designed to perform a specific function as indicated in Table 2-1.

Table 2-1. Component Functions.

Grass buffer strip	Reduces runoff velocity and filters particulates from runoff.
Sand bed	Reduces velocity, filters particulates, and spreads flow over the length of the bioretention area. Aeration and drainage of the planting soil are provided by the 18 in. (0.5 m) deep sand bed.
Ponding area	Provides temporary storage for runoff prior to evaporation or infiltration. Particulates not filtered out in the grass filter strip or the sand bed settle within the ponding area.
Organic/mulch layer	Filters pollutants and provides an environment conducive to growth of microorganisms, which degrade petroleum-based products and other organic material. This layer acts in a manner similar to the leaf litter in a forest, preventing the erosion and drying of the underlying soils. The maximum sheet flow velocity before erosive conditions is 0.9 m/s (3 ft/s).
Planted ground cover	Reduces the potential for erosion slightly more effectively than mulch. The maximum sheet flow velocity prior to erosive conditions is 0.3 m/s (1 ft/s).
Clay in planting soil	Provides adsorption sites for hydrocarbons, heavy metals, nutrients, and other pollutants.
Voids in planting soil	Provide stormwater storage, making the stored water and the nutrients in the water and soil available for uptake by plants.

The basic bioretention area can be customized to accommodate specific needs. For example, the City of Alexandria, Virginia, modified the bioretention BMP because impervious subsoils and marine clays prevented complete infiltration into the soil system. The redesigned bioretention BMP, which includes an underdrain bed that collects the infiltrated water and discharges it to a downstream sewer system, acts more as a filter that discharges treated water than as an infiltration device. Design modifications that may include both aerobic and anaerobic zones in the treatment area are being evaluated. The anaerobic zone will promote denitrification.

2.12.1 Research Needs

Bioretention is an emerging BMP that needs research into almost all phases of design, operation, maintenance, performance, and cost. However, since the system consists mainly of landscaping, capital costs can be estimated fairly easily. On the basis of the literature review, the following research needs were identified:

- Sizing of the bioretention area;
- Potential impacts of various pollutant loads on underlying soil and groundwater;
- Impacts of various sand and mulch layer depths on water quality; and
- Operation and maintenance costs.

2.13 Porous/Pervious Paving

Porous paving as used in the United Kingdom is used for light vehicle loading in car parks and accesses. The term describes a system that consists of a load-bearing, durable surface overlaying a layered structure that temporarily stores water before it infiltrates or drains to a controlled outlet. A typical porous pavement is shown in Figure 2-18. The surface (e.g., grass and gravel, porous concrete, or porous asphalt), itself can be porous allowing water to infiltrate across it, or it can be built of impermeable blocks separated by spaces and joints, through which the water can drain. The latter system is termed permeable paving. Attenuation of flow is provided by storage within the underlying structure or sub-base, together with appropriate flow controls. A layer of geotextile may be used to permit groundwater recharge, thus helping to restore the natural water cycle. Alternatively, where infiltration is inappropriate (e.g., where the groundwater quality is particularly vulnerable or where the soil type is unsuitable), the surface can be constructed above an impermeable membrane. The system serves as a convenient means of drainage in spatially constrained urban areas.

Porous pavements can be used to achieve significant attenuation of flow and improvement in water quality. The surface and subsurface infrastructure can remove both soluble and particulate pollutants in the urban runoff. Roof drainage can be piped directly into the storage area. Also, lined systems make it possible to pipe the stored runoff out of reuse.

Figure 2-18. Typical Porous Pavement Installation.

2.13.1 Research Needs

On the basis of the literature review, the following research needs were identified for porous pavement:

- Performance of saturated subgrades;
- Aggregate grading and performance under load;
- Performance of geosynthetic structures;
- Standard specifications for materials and installation;
- Methods of block replacement;
- Effects of vertical and horizontal attenuation processes through the granular subfill and by evaporation;
- Water quality aspects of porous pavements, including the effects of polluted runoff, seasonal changes in pollutant loadings and throughput, and the impact of pollutants on soil and groundwater;
- Specific research on geotextiles:
  - Should a geotextile be installed at the high level, below the surfacing materials, or is it better to accumulate sludge at the base of pavement?
  - If geotextile is installed at one or more locations, which types and characteristics would best enhance its pollutant-trapping efficiency?
- Length of useful life;
- Effects of operation and maintenance on whole-life performance;
- Effectiveness of various operation and maintenance techniques and equipment;
- Effectiveness of various maintenance intervals; and
- Installation costs.

CHAPTER 3.0

CURRENT STATUS OF BMPs/SUDS

3.1 Introduction

Water quality issues, such as protection of receiving waterbody ecology and drinking water supplies, have forced changes in stormwater management. Many local governments in the United States and United Kingdom have implemented structural means to control both the quantity and the quality of urban stormwater runoff at or near to its source. These means are called best management practices—BMPs in the United States—and sustainable urban drainage systems—SUDS in the United Kingdom.

Conventional stormwater controls rely on drainage channels and underground pipe systems to prevent flooding by conveying the water away as quickly as possible. Most of the conventional drainage systems are not effective in removing the pollutants carried in stormwater. Also, conventional systems tend to ignore amenities such as landscaping potential and the provision of wildlife habitat.

According to the Construction Industry Research and Information Association (CIRIA), SUDS are more sustainable than conventional means of drainage because they integrate the following features:

- Manage runoff flow rates, reducing the flooding resulting from urbanization;
- Protect or enhance water quality;
- Are compatible with the environmental setting and the needs of the local community;
- Provide a habitat for wildlife in or along urban watercourses; and
- Encourage natural groundwater recharge (where appropriate).

In the United States, BMPs tend to focus on water quality rather than quantity, primarily because EPA has no authority in the area of flood control despite the obvious adverse environmental impacts of the larger volumes and higher rates of runoff brought about by urbanization. Also, BMP programs in most communities do not emphasize amenities such as improving the environmental setting and providing wildlife habitat.

This chapter describes the primary regulatory programs and current status of BMPs and SUDS in both the United States and the United Kingdom.

3.2 BMP Programs in the United States

The widespread use of BMPs in the United States is primarily the result of various federal, state, and local regulatory programs.

3.2.1 Federal Programs

The implementation of most BMPs has been driven by two components of the 1972 Clean Water Act (CWA), which was amended in 1987 to include stormwater discharges. This act requires states to assess the condition of the surface waters within their jurisdiction to determine whether they are "fishable and swimmable." Where water quality is not high enough to sustain these beneficial uses, a total maximum daily load (TMDL) must be developed for the constituents that are contributing to the impairment of water quality.

The TMDL is the maximum pollutant load (flow times pollutant concentration) that can be assimilated by a waterbody without impairing its beneficial uses. More than 20,000 such impaired waters have been identified throughout the states, constituting more than 300,000 miles of rivers and streams and more than 5 million acres of lakes. Once a TMDL for a specific pollutant has been developed for a particular surface water body, the load is allocated to each discharger to that waterbody. As TMDLs are developed for the impaired segments, dischargers, including municipalities, may have to implement stormwater BMPs to reduce their contribution of the pollutants that cause the impairment.

Under CWA, municipalities must obtain National Pollutant Discharge Elimination System (NPDES) permits for stormwater discharges. The NPDES permit requires municipalities to develop stormwater management plans, which involve implementation of BMPs for controlling pollutants in stormwater runoff. Because implementation of BMPs is driven almost exclusively by elements of CWA, reduction of the pollutant loads is the near-exclusive objective of implementation, with little regard to either quantity control or amenity/habitat value which are key features of SUDS. It should be noted, however, that a large number of stormwater detention basins were built because of local and state needs for hydraulic controls before CWA.

The NPDES stormwater program was implemented in two phases:

3.2.1.1 Phase I of NPDES Stormwater Program

In November of 1990, EPA promulgated the first phase of the regulations governing the application for NPDES stormwater permits. Phase I required that large municipal separate storm sewer systems serving populations greater than 100,000 (MS4) and some industrial facilities prepare stormwater management and pollution prevention plans as part of their NPDES permit application.

MS4 Phase I permits effectively prohibited nonstormwater discharges into the storm sewers and required that the discharge of pollutants be reduced to the maximum extent practicable. These objectives were achieved through BMP control techniques, system design and engineering methods, and other provisions appropriate for the control of such pollutants.

The industrial permitees of Phase I included facilities covered by federal effluent guidelines as well as those in other designated classes of industry and commerce, such as those engaged in hazardous waste treatment, storage, and disposal; landfills; recycling; vehicle maintenance and equipment cleaning; sewage sludge handling; construction activity (sites that include 5 or more acres of disturbed land); and facilities containing materials exposed to stormwater. Permittees were required to prepare a stormwater pollution prevention plan that described pollution sources, measures, and controls using BMPs.

3.2.1.2 Phase II of the NPDES Stormwater Program

On January 9, 1998, EPA proposed NPDES regulations for the second phase of stormwater discharge control for communities with a population less than 100,000 and generally greater than 50,000. The final Phase II rule was published on December 8, 1999. Under this phase, NPDES permits are required by the end of 2002 for stormwater discharges from MS4s and construction sites that contain between 1 and 5 acres of disturbed land.

Phase II requirements are heavily dependent on the use of BMPs.

To obtain a permit, the facility must develop and implement a stormwater management program that addresses six minimum control measures:

- Public education and outreach on stormwater impacts;
- Public involvement/participation;
- Detection and elimination of illicit discharges;
- Control of stormwater runoff from construction sites;
- Post-construction stormwater management in new developments and redevelopments; and
- Pollution prevention/good housekeeping for municipal operations.

The minimum control, post-construction stormwater management in new developments and redevelopments, has resulted in much wider use of BMPs for urban drainage systems in the United States.

3.2.1.3 Endangered Species Act

Independently of CWA, the Endangered Species Act (ESA) of 1973 also created opportunities for implementing BMPs. The ESA protects plants and animals that are currently in danger of extinction (endangered) and those that may become endangered in the foreseeable future (threatened). BMPs are used to protect the habitat of these species by both improving the quality of water and reducing the quantity of stormwater runoff.

The listing of several salmon stocks under the ESA in 1999 highlighted the importance and impact of stormwater runoff on salmon and other species of fish. The endangerment of salmon, a keystone species, raised concerns that additional species may become threatened.

In response to the endangerment of salmon, the Washington State Department of Transportation (WSDOT), the Washington Department of Ecology (DOE), and many other local organizations have made programmatic changes in regard to stormwater management. The WSDOT has updated its Highway Runoff Manual to reflect regulatory changes that apply to all areas where salmon are listed as threatened or endangered species. The Washington DOE has updated its *Technical Guide for Stormwater Management* to reflect current science and regulations, and is evaluating the need for a comprehensive stormwater management plan.

3.2.2 State and Local Stormwater Programs

Many states have developed stormwater management statutes to satisfy the federal requirements and their unique hydrogeological conditions. These statutes may include specific requirements for individual counties or cities. For example, the State of Maryland requires the BMPs for some counties to be designed to accommodate the 2-year storm, whereas those of other counties must be capable of handling the 10-year storm.

Some states require counties to establish their own stormwater management criteria based on the counties' individual hydrologic concerns. In Maryland, the county has the primary responsibility to maintain interagency coordination, to provide annual reporting to the state, to establish source identification (including discharge detection and elimination), and to provide public outreach plans and watershed restoration plans to the state.

The City of Orlando, Florida, and the neighboring counties have a large number of natural lakes and ponds. Most residents are never far from a body of high-quality, high-amenity open water. Their drinking water is abstracted from an aquifer below the city. This means that the ponds and the watersheds that contribute to groundwater recharge must be protected from pollution. BMPs in Orlando include open ponds and grassed detention/infiltration basins. Relatively costly solutions have been developed to target water quality, particularly downtown where the land values are high.

Austin, Texas, is located in a hot and humid climate where annual area rainfall averages approximately 800 mm (31 in.). There are a number of rivers in the area, but the community has been relying on a spring from the Edwards Aquifer, which yields very high-quality water that is valued by the residents as a recreation site. Consequently, the city has developed strict regulations that require the installation of sand filters or equivalent, generally wet ponds, to attenuate and treat rainwater from every development. Highways are state facilities, which are exempt from city regulations, but are subject to state rules that require treatment for facilities located in the Edwards Aquifer. The regulation, Austin 2002, is very proscriptive, giving detailed sizing guidance that has resulted in the construction of concrete-walled structures, which are expensive to construct and maintain. Despite the substantial cost of BMP development compared to the costs incurred by other cities in Texas, the regulation is well supported by the community so that the quality of the groundwater is not compromised. Thousands of BMPs can be found within the city limits. In addition, the amount of impervious cover in some environmentally sensitive watersheds is limited to as little as 15% of the site area as an additional measure of protection.

3.2.3 Chesapeake Bay Programs

Many environmental awareness organizations in the United States have implemented stormwater BMPs. The Chesapeake Bay Program is a prominent example of the use of BMPs in a stormwater management plan implemented by such an organization.

The Chesapeake Bay, the largest estuary in the United States and one of the most productive in the world, was the nation's first estuary targeted for restoration and protection by nonregulatory action. Water quality in the bay had been on the decline for nearly 40 years, as evidenced by widespread loss of submerged aquatic vegetation and by periodic fishkills. The Chesapeake Bay Program was developed as the means to restore this valuable resource.

The Chesapeake Bay Program, begun in 1985, is a compact of state and federal governments to restore the Bay's water quality. In 1987, the States of Maryland, Virginia, and Pennsylvania; the District of Columbia; and EPA set a goal of reducing nutrient releases to the bay by 40% by the year 2000.

The establishment of these goals resulted in the development of several new programs. For example, Virginia's response was the Chesapeake Bay Preservation Act, also known as the Bay Act, adopted by the Virginia General Assembly in 1988. Although the Chesapeake Bay Program established "goals" for the Chesapeake Bay estuary, the Bay Act established the regulatory means by which to attain these goals.

Like the Chesapeake Bay Program, the Bay Act helped to generate many other programs, for example, the Chesapeake Bay Local Assistance Department (CBLAD), which helped implement the Bay Act. CBLAD is headed by a nine-member citizen board representing various stakeholder interests—such as business, agriculture, land development, local government, and environmental protection—as well as the various geographic regions within Virginia. The board, stakeholders, and regulatory agencies work together toward the goals set by the Chesapeake Bay Program. Their work has resulted in the application of BMPs throughout the bay region of Virginia.

3.2.4 BMPs and Water Supply Utilities

Water utilities are beginning to use BMPs as a part of their source water management programs. The health of aquatic ecosystems is a concern for water utilities that rely on surface water sources. Sediments, nutrients, pesticides, and pathogens in source water reduce treatability, increase treatment costs, and ultimately increase the risk to public health. Until recently, utilities responded to more stringent regulations by increasing chemical dosages or by altering their treatment processes. However, today they are considering the use of BMPs in the watershed to preserve or improve the quality of their raw water supply.

3.2.5 Low-Impact Development and Smart Growth

Demographic shifts and economic and environmental concerns have spurred the Smart Growth movement. BMPs incorporated into better site design are being used as preventive measures.

Across the nation, many states are exploring Smart Growth initiatives. State and local planners have recognized the excellent opportunity Smart Growth offers to reduce stormwater runoff and pollutant export. The amount of impervious cover is the key variable in influencing the amount of nutrients delivered in stormwater runoff. Low impact development incorporates more open space into site design. Coupled with BMPs, low-impact development has a promising future.

3.3 SUDS in the United Kingdom

This section presents a brief overview of the development and current status of SUDS in the United Kingdom. The various technical, political, social, and economic factors discussed below illustrate the legislative complexity of drainage matters in the United Kingdom. Although this situation is by no means unique—similar complexity is found in many other countries—it does show why progress towards implementation has been slow, despite the wide acceptance of the underlying principles.

SUDS are a relatively new concept in urban drainage in the United Kingdom, and the concepts and methodologies are still being developed. Progress in Scotland is ahead of the progress in England and Wales, and all planning applications for new developments in Scotland must now incorporate SUDS in their drainage strategies and designs. A great deal of public education and technical training has been required, and a considerable amount of research has been commissioned to investigate system performance, maintenance, and the associated costs. The general benefits of SUDS in the United Kingdom are described on CIRIA's Website (November 2001):

> Drainage systems can be developed in line with the ideals of sustainable development by balancing the different issues that should be influencing the design. Surface water drainage methods that take account of quantity, quality, and amenity issues are collectively referred to as Sustainable Urban Drainage Systems (SUDS). Systems are more sustainable than conventional drainage methods for these reasons:
>
> - They manage runoff flow rates, reducing the flooding resulting from urbanization.
> - They protect or enhance water quality.
> - They are supportive of the environmental setting and the needs of the local community.
> - They provide a habitat for wildlife in and along urban watercourses.
> - They encourage natural groundwater recharge (where appropriate).
>
> They do this by the following means:
>
> - Dealing with runoff close to where the rain falls.
> - Managing potential pollution at its source now and in the future.
> - Protecting water resources against pollution from point sources (such as accidental spills) and diffuse sources.
>
> They may also allow new development in areas where existing

sewerage systems are close to full capacity, thereby enabling development within existing urban areas.

Urban drainage is moving away from the conventional thinking of designing for flooding and the adverse impacts of conventional piped surface water sewerage systems to balancing the impact of urban drainage on flood control, quality management, and amenity.

More details of the current work on SUDS in the United Kingdom can be found at CIRIA's Website: www.ciria.org.uk/suds.

3.3.1 Legislative Background

Legislation affecting drainage in the United Kingdom is complex. Because sustainable urban drainage techniques were not widely practiced in England, Wales, Scotland, or Northern Ireland when the legislation was passed, they are not dealt with explicitly. The planning and approval processes as they currently apply in England, Wales, and Scotland are presented in Appendix B.

The pending implementation of the Water Framework Directive in Europe has led to a shift in thinking towards more ecological objectives in planning river systems. The European Union nations must have major initiatives in place by 2004, in particular, those dealing with watercourse degradation caused by diffuse pollution. In the United Kingdom, there has been significant investment in sewage treatment infrastructure during the last 15 years as the key element in meeting water quality targets. The Water Framework Directive will influence future thinking on drainage matters, and SUDS are expected to take an increasingly important role although it must be acknowledged that it may take several years for SUDS to play a key role.

The concept of the treatment train has been developed to provide a hierarchical basis to the development of practical means of addressing diffuse pollution. Using the concept of the treatment train, a variety of approaches has been developed to mitigate the effects of urban runoff. The concepts are currently being applied to agricultural practices, and in the future they will be extended to silviculture.

3.3.2 England and Wales

The environmental regulator (the Environment Agency) is responsible for flood protection, which accounts for a substantial fraction of its spending. Although the agency is empowered to deal with the water quality issues arising from diffuse pollution, until recently this has not been a high priority issue, and SUDS have been implemented only in a few locations.

In England and Wales, the key stakeholders in the management of the water cycle are the water companies. These private utility companies are strictly regulated by the Office of Water Services (Ofwat) and the Environment Agency, which is actively promoting the implementation of SUDS in all new developments, and a SUDS working group has published a document, *Framework for Sustainable Drainage Systems (SUDS) in England and Wales*. This is undergoing a period of widespread consultation which concluded on August 15, 2003. Although the water companies are supportive of the principles behind SUDS, they are reluctant to wholeheartedly accept SUDS until certain outstanding ownership and adoption issues (beyond the scope of this study) are resolved. The water companies also believe that adoption of SUDS will lead to additional responsibilities and costs that will not be recoverable under the current regulatory pricing regimen.

During a meeting with the project team, Ofwat confirmed that it supported the principles behind SUDS and would have no objections to the water companies using these techniques, where technically appropriate. However, Ofwat did recognize that an agreement had to be reached between it and the water companies concerning the recovery of costs allowable through the pricing regimen.

3.3.3 SUDS in Scotland

Scotland already has numerous SUDS in place. Implementation of SUDS has been driven by water quality considerations and the need to improve the status of small watercourses, many of which are close to urban areas. Scotland has made good progress in meeting the objectives of the Water Framework Directive, and there is good collaboration between the regulators and the developers and industry. A clear environmental policy, which has been in place for several years, has resulted in general design guidance (CIRIA, 2000a), and most developers see the benefits of a "green" environment. Because the choice of the type of SUDS in Scotland rests with the local government, a wide variety of different types of systems have been installed, some of which are discussed below.

The widespread use of SUDS in Scotland is primarily the result of the Scottish Environment Protection Agency (SEPA) and its predecessor, the Forth River Purification Board. At the end of 2001, there were 3913 SUDS components on 767 sites around Scotland (Wild, Jeffries, and D'Arcy, 2002). The SEPA policy has been directed primarily towards alleviating diffuse pollution, but has not been specific regarding the particular types of SUDS that should be implemented; rather, the emphasis has been on following a hierarchy, termed the treatment train (Figure 3-1).

The successful passage through the Scottish Parliament of the Water Environment and Water Services (Scotland) Act 2003 on 29th January 2003 amended the Sewerage (Scotland) Act 1968. To facilitate the implementation of the SUDS policy in Scotland, a Drainage Impact Assessment (DIA) procedure was adopted. DIA provides a coherent format and procedure for the assessment of impacts and delivery of solutions through the statutory processes, which might involve several authorities. The DIA procedure encompasses disposal of foul and surface water flows, together with the assessment of flood risks.

Initially, a set of principles were published with very few design parameters (CIRIA; 2000a); design and details have become the responsibility of the developer. Figure 3-2 shows the numbers of different SUDS components in use in Scotland at the end of 2001. On most sites, a single type of SUDS component had been installed, and it is apparent that in Scotland the stormwater treatment train approach to urban drainage has not yet become widely accepted.

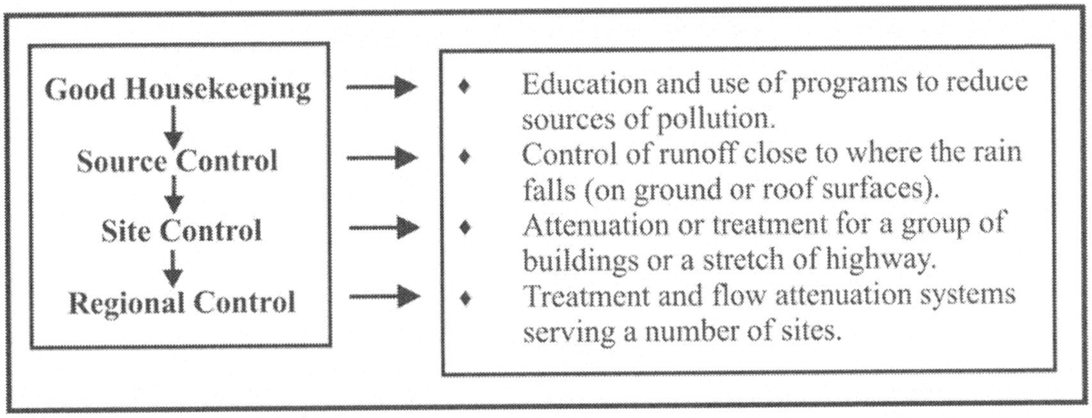

Figure 3-1. The Stormwater Treatment Train.

Although there are considerable numbers of swales, detention basins, and retention ponds, the predominant type of SUDS are in-ground and pipe based, that rely on settling and filtration of contaminants, rather than on exfiltration to groundwater. This probably reflects the change from earlier methods and the small scale of most developments as well as the belief that in-ground systems are particularly reliable or inexpensive. Because these systems do not offer any amenity benefits, it is particularly interesting to see that significant numbers of retention ponds and wetlands have been constructed. It is to be expected that retention ponds and wetlands would be relatively uncommon, because according to national guidance (CIRIA, 2000a), they should be used primarily as regional controls in especially large developments and at sites where there is the potential for severe pollution.

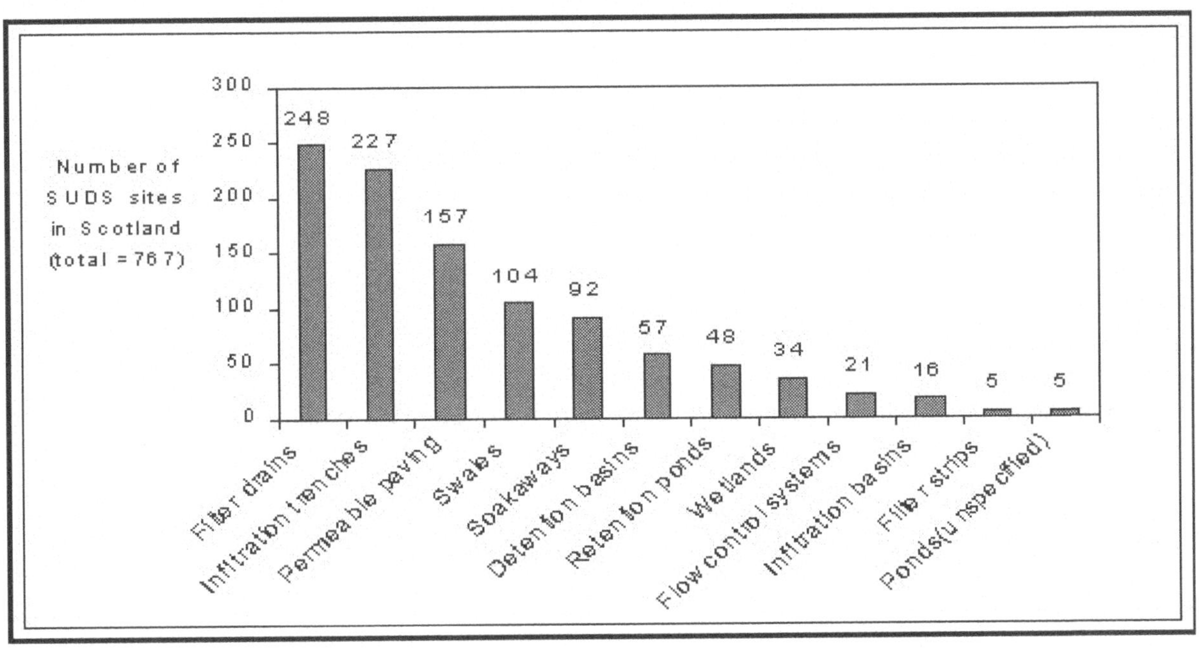

Figure 3-2. Numbers of SUDS sites in Scotland (01/01/2002) (Wild, Jeffries, and D'Arcy, 2002).

A wide variety of SUDS are available for enhancing the ecological benefits of stream systems in urban environments. Ideally the design of such systems should be based on the concept of the treatment train. In reality, it is likely that the treatment train concept will be used only in larger applications, and in most cases, smaller, site-based systems will be used. In Scotland, only 26% of the SUDS installations were found to employ more than one stage of treatment.

The requirement for developers in Scotland to utilize the treatment train approach is not very stringent. The Planning Advice requires merely that SUDS be selected and designed in accordance with CIRIA's *Manual* (CIRIA, 2000a). Although this design manual was developed for use in Scotland, it gives only general guidance, which has resulted in a wide range of types of systems being installed, some of which are unsuitable for their locations. In contrast, where there is detailed guidance, adoption of SUDS has become widespread.

In the United States, the guidelines for BMPs are more clearly defined, and there is little or no debate over the relative merits of different types of systems. It is unlikely that the use of the stormwater management train will become standard practice in Scotland until a more detailed and specific guidance is developed, with more responsibilities for adoption and maintenance of SUDS. This may require a new legal framework through appropriate legislation.

3.3.4 Other U.K. Issues

Two additional concerns that are impeding the implementation of SUDS in the United Kingdom are the separation of responsibilities of the highway and drainage authorities and the need for education of those charged with implementing more sustainable drainage approaches. Traditionally, highway engineers have produced designs that direct water away from highways, whereas sustainable drainage requires disposal of the runoff from rainfall and snowmelt closer to the point where it is generated (road or roof). Designs that address this problem are essential, but highway engineers are reluctant to adopt this approach. The most successful solutions are those that use a cooperative approach which integrates all disciplines and produces clear and unambiguous guidelines for new system components.

3.3.5 Research and Development

Various research and development (R&D) projects into SUDS are in progress in the United Kingdom. A summary of this work is given in Appendix C, provided through the courtesy of CIRIA.

CHAPTER 4.0

PERFORMANCE PROTOCOLS

4.1 Background

Many studies have assessed the ability of stormwater BMP/SUDS to reduce pollutant concentration and loadings in stormwater. Although some of the monitoring projects conducted to date have done an excellent job of describing the effectiveness of specific BMP/SUDSs and BMP/SUDS systems, there is a lack of standards and protocols for conducting consistent BMP/SUDS assessment and monitoring. This lack becomes readily apparent when seeking to summarize the information gathered from the various BMP/SUDS evaluations, as in this study. Inconsistent methods, lack of associated design information, and the differing reporting protocols make wide-scale assessments difficult, if not impossible (Strecker and Urbonas, 2001; Urbonas, 1994). For example, individual studies often include the analysis of different constituents, utilize different methods of data collection and analysis, and report varying amounts of information on BMP/SUDS design and flow characteristics. The differences in monitoring methods and data evaluation alone contribute significantly to the range of BMP/SUDS "efficiency" that has been reported in literature to date.

In April 2002, the American Society of Civil Engineers (ASCE) in association with EPA published a guidance manual, *Urban Stormwater BMP Performance Monitoring: A Guidance Manual for Meeting the National Stormwater BMP Database Requirements*. The document was prepared by GeoSyntec Consultants, Urban Drainage and Flood Control District, and the Urban Water Resources Research Council of the ASCE. Sections of the document were developed concurrently with the Federal Highway Administration's *Guidance Manual for Monitoring Highway Runoff Water Quality*, and material was also adapted from the Washington State Department of Ecology's *Stormwater Monitoring Guidance Manual*, published in November 1995.

The ASCE manual is intended to improve the state of the practice by providing a recommended set of protocols and standards for collecting, storing, analyzing, and reporting BMP monitoring data that will lead to better understanding of the function, efficiency, and design of urban stormwater BMPs.

There are no protocols or guidance on the monitoring of SUDS in the United Kingdom.

A document similar to the ASCE manual would significantly improve the value of the information being compiled in the United Kingdom for sustainable drainage systems, would ensure the availability of sound scientific data for determining which systems function most effectively under what conditions, and to what degree design and environmental variables directly affect BMP/SUDS efficiency.

Some of the material that follows is taken directly from the 2002 ASCE publication, which ensures that this study reflects the latest U.S. recommendations and thinking in this area. It is, however, recognized that the manual focuses primarily on the collection, reporting, and analysis of water quantity and quality at the heart of quantitative BMP efficiency projects. It does not address any qualitative monitoring that may often be equally useful for measuring and monitoring water quality for assessing BMP/SUDS efficiency.

4.2 Proposed Approach

Although quantitative monitoring provides essential evidence of BMP/SUDS performance in managing both hydraulic and chemical loads of stormwater runoff, it cannot address "softer" performance issues, such as amenity objectives. In addition, the prohibitive costs of quantitative monitoring generally rule out the use of such methods for establishing the knowledge base required for these systems.

Many BMP/SUDSs, such as wet ponds, have been the subject of numerous studies of their pollutant removal ability; therefore, little additional knowledge would be gained by further monitoring of the same type. Consequently, it can be assumed that facilities designed using the same guidelines and properly maintained will achieve similar performance.

Because the objective of this study is the establishment of the cost and maintenance requirements of these facilities, empirical observation of conditions at these sites is necessary to develop the needed information. Empirical observations on the frequency of maintenance, the locations and amounts of sediment accumulation, the effects of trash and other debris on outlet clogging (detention basins), the effects of vegetation height and density on flow patterns (swales), and similar factors will provide valuable information that is usually not considered in studies that focus on changes in runoff quality.

A combination of quantitative monitoring to fill in the gaps in our understanding and qualitative analyses to expand our knowledge is proposed as the best approach during Phase 2 of this project.

4.3 Performance Monitoring Objectives

Studies evaluating the efficiency of BMP/SUDS in attaining water quality goals should address the following questions:

- What degree of pollution control or improvement in effluent quality does the BMP provide under normal conditions?
- How does this efficiency vary from pollutant to pollutant?

- How does this normal efficiency vary with the magnitude of storm events?
- How does this normal efficiency vary with rainfall intensity?
- How does this efficiency vary with antecedent conditions?
- How does this efficiency vary with various input concentrations?
- How do design variables affect efficiency?
- How does efficiency vary with different operation and/or maintenance approaches?
- Does efficiency improve, decline, or remain stable over time?
- How does this system's efficiency compare with the efficiency of other systems?
- Does the system cause an improvement in or protect downstream biotic communities?

Studies that evaluate the efficiency of BMP/SUDS in attaining flood alleviation should address the following questions:

- Are peak flow rates from the developed areas reduced?
- Are peak flow rates reduced to predevelopment levels?
- How does the extended duration of higher flows (following detention to reduce the peak) affect downstream channel morphology?
- Are the total volumes discharged into the receiving watercourse reduced?
- Are floodwaters diverted appropriately once the system capacity is exceeded?

Studies that evaluate the efficiency of BMP/SUDS in attaining aesthetic, ecological, and amenity goals should address the following questions:

- Does the project improve the appearance of a site?
- Does the project improve wildlife habitat and biodiversity of the site?
- Does the project result in enhanced amenities to the local community?

4.4 Quantitative Monitoring

4.4.1 Challenges of Quantitative Monitoring

Information collected on the efficiency BMP/SUDS serves a variety of goals and objectives. However, the principal challenge is the great temporal and spatial variability of stormwater flows and pollutant concentrations. Stormwater quality at a given location varies greatly both during a single storm event and between storms; thus a few samples are not likely to be representative of stormwater quality at a given site or of its effect on a given system. Therefore, an extensive sampling program is generally needed to answer many of the questions regarding BMP efficiency with a high level of statistical confidence. A poorly designed monitoring program could lead to erroneous conclusions.

The following options could be considered when attempting to reduce the cost of quantitative monitoring:

- Limiting the system types to be considered to those where only a few prior studies have been conducted; and
- Limiting the number of constituents evaluated.

The task of summarizing or comparing information from several sites is further complicated by the multitude of site variables including the following:

- BMP design, e.g., length, width, height, storage volume, outlet design, etc.
- Influent runoff concentrations
- Geographical location
- Watershed size
- Percent imperviousness
- Vegetative canopy
- Soil type
- Watershed slopes
- Compaction of soils.

Climatic variables that may have a strong influence on system performance include the following:

- Rainfall intensity
- Flow rate
- Season
- Vegetation
- Inter-event timing

The extent of maintenance is another variable that is likely to exert significant influence on system performance.

4.4.2 Parameters for Monitoring

The U.S. Nationwide Urban Runoff Program (NURP) has adopted consistent data collection techniques and analytic parameters for meaningful comparison of the gathered data. The following constituents are listed by NURP as "standard pollutants characterizing urban runoff" (U.S. EPA, 1983):

SSC = Suspended solids concentration
BOD = Biochemical oxygen demand
COD = Chemical oxygen demand
Cu = Copper
Pb = Lead
Zn = Zinc
TP = Total phosphorus
SP = Soluble phosphorus
TKN = Total Kjeldahl nitrogen
$NO_2 + NO_3$ = Nitrate + Nitrite

The following factors were considered when including a pollutant in the list of constituents recommended for monitoring (Strecker, 1994):

- The pollutant has been identified as prevalent in typical urban stormwater at concentrations that could cause water quality impairment.

- The analytical test used can be related to potential water quality impairment.
- Sampling methods for the pollutant are straightforward and reliable for a moderately careful investigator.
- Analysis for the pollutant is economical on a widespread basis.
- Treatment is a viable option for reducing the load of the pollutant.

Table 4-1 shows the most basic arrangement of constituents that should be considered for monitoring.

Table 4-1. Typical Urban Stormwater Runoff Constituents and Recommended Detection Limits (ASCE, 2002).

Constituent	Units	Target Detection Limit
Conventional		
pH	pH	N/A
Turbidity	mg/L	4
Total Suspended Solids	mg/L	4
Total Hardness	mg/L	5
Chloride	mg/L	1
Bacteria		
Fecal Coliform	MPN/100 mL	2
Total Coliform	MPN/100 mL	2
Enterococci	MPN/100 mL	2
Nutrients		
Orthophosphate	mg/L	0.05
Phosphorus-Total	mg/L	0.05
Total Kjeldahl Nitrogen (TKN)	mg/L	0.3
Nitrate–N	mg/L	0.1
Metals – Total Recoverable		
Total Recoverable Digestion	µg/L	0.2
Cadmium	µg/L	1
Copper	µg/L	1
Lead	µg/L	1
Zinc		
Metals – Dissolved		
Filtration/Digestion	µg/L	0.2
Cadmium	µg/L	1
Copper	µg/L	1
Lead	µg/L	5
Zinc		
Organics		
Organophosphate Pesticides (scan)	µg/L	0.05 – 0.2

Different forms (species) of metal have different levels of toxic effects. In general, metals are most toxic in their dissolved, or free ionic, form. The distribution of pollutants between the dissolved and particulate phases will depend on where in the system the sample is collected. Hardness values for receiving waters are often different from those for stormwater. Hardness affects the bioavailability of heavy metals, further complicating their ecological impact in dissolved form.

Total recoverable metals data together with dissolved metals data can be used to assess potential metals sediment issues.

4.4.3 Period of Analysis

The period of analysis used in a monitoring program is important. It should take into account how the constituent of interest changes over time, allowing observation of relevant changes in the efficiency of the SUDS/BMP on the time scale in which such changes occur. For example, the effluent quality in terms of nutrients improves in a wetland during the growing season. The opposite is true during the winter months or whenever decaying litter and plant material contribute significant quantities of nutrients and, potentially, other contaminants. Therefore, monitoring results may need to be analyzed differently during different seasons. Some studies have also reported that nutrient removal in wet ponds is reduced once the vegetation becomes established; consequently, unless vegetation harvesting is part of routine (annual) maintenance, nutrient monitoring should focus on well-established systems. Monitoring such variations in performance, and thus efficiency, on a temporal scale is extremely important to understanding how a specific BMP functions.

In addition to observing how factors such as climate affect SUDS/BMP efficiency as a function of time, it is important to relate the monitoring period to the potential impact a given constituent would have on the receiving water. For example, it may not be useful to study the removal of some heavy metals, such as mercury, for a short period since the adverse impacts of such contaminants generally become evident over long-term accumulation in sediments and biota. Likewise, some temperature, BOD, DO, pH, TSS, and metals concentrations may have a significant impact in the near term.

Table 4-2. Examples of Water Quality Characteristics and Relevant Monitoring Periods.

Time Scale for Analysis	Water Quality Constituent
Short-term	BOD, DO
Long-term	Organics, Carcinogens
Short-term and long-term	Metals, TSS, Nitrogen, Phosphorus, Temperature, pH, Pesticides

4.4.4 Performance Efficiency Computations: Concentrations, Pollutant Loads and Event Mean Concentration Monitoring

There is much variation and disagreement in the literature about what measure of efficiency is best applied in specific situations; however, it is generally accepted that event mean concentrations and long-term loadings are the best indicators of the effects of the SUDS/BMP on acute and chronic pollution, respectively.

Various methods are available for assessing and quantifying the amounts of pollutants conveyed to and from a SUDS/BMP. Three measures are most commonly used:

- Concentrations of stormwater at a given time;
- The total load conveyed over a specified duration; and
- The event mean concentration (EMC).

4.4.4.1 Concentrations

Concentrations measured at a given time can be useful for evaluating BMP/SUDS efficiency in a number of ways. Concentrations in samples collected at specific times during an event can be used to generate pollutographs, i.e., a plot of the concentration of pollutants as a function of time. The pollutographs facilitate the analysis of intra-event temporal variations in runoff concentration, for example, for determining whether the "first-flush" phenomenon was observed for a specific event. Detailed concentration data are one of the means of assessing the concentrations of pollutants that have acutely toxic effects, particularly where runoff from storm events constitutes a significant proportion of downstream flow. Under some circumstances, reduction of peak effluent concentrations may be more important than reduction of the event mean concentration. Implementing a monitoring program that collects sufficient data to evaluate the temporal variation in runoff and BMP/SUDS effluent concentration can be costly. The trade-off between collecting data from a larger number of events and collecting detailed concentration data from intra-storm periods often limits the utility of studies that collect detailed concentration data. This type of detailed monitoring is best focused on monitoring only the outflow rather than monitoring both inflow and outflow.

4.4.4.2 Loads

Loads are typically calculated by physically or mathematically combining a number of individual concentration measurements which have been assigned an associated flow volume. A variety of methods can be used for estimation of loads, depending on the sampling and flow measurement techniques used. Sampling methods include timed samples, flow weighted samples, or combinations of both. Flow measurements can be taken continuously or intermittently, or can be modeled from other hydrologic information such as the readings from rain gauges or gauges in a nearby watershed. Many BMP/SUDS monitoring studies focus on sampling only the water quality and neglect flow measurement. Accurate flow measurement or well-calibrated flow modeling is essential to determination of pollutant loadings.

Data on pollutant loadings are often most useful in assessing the effectiveness of a BMP/SUDS for lakes or estuaries where long-term loadings can cause water quality problems outside storm events. Where the flow rate from a particular BMP/SUDS is small in comparison with the flow rate in the receiving waterbody, downstream impairments are typically dependent not on concentrations, but on the absolute load of pollutant reaching the receiving water. For example, pollutant loads are the central issue in BMP studies associated with receiving waterbodies that are regulated under the Total Maximum Daily Load (TMDL) program, particularly where pollutants deposited in slow-moving systems are of concern. Dry weather flows can also contribute substantially to long-term loadings. In addition, "on-line" BMP/SUDS (ponds) that receive appreciable amounts of dry weather flows may have reduced "capacity" for storage of wet weather pollutants. For example, a pond's performance may be affected by the amount of water it contains before the event.

4.4.4.3 Event Mean Concentrations

Event mean concentration (EMC) is a statistical parameter used to represent the flow-proportional average concentration of a given constituent during a storm event. It is defined as the total constituent mass divided by the total runoff volume. When combined with flow measurement data, the EMC can be used to estimate the pollutant loading from a given storm. Using the EMC approach to understand BMP efficiency is applicable primarily to wet weather flows. In most cases, the EMC is the most useful means of quantifying the amount of pollution caused by a runoff event. Collection of EMC data has been the primary goal of the National Stormwater Best Management Practices Database Project.

4.4.5 Methods of Analysis

The following method is recommended by the ASCE manual for use in analyzing data from both existing and new monitoring studies.

4.4.5.1 Effluent Probability Method

The most useful approach to quantifying BMP/SUDS efficiency is to determine first whether the BMP/SUDS is providing treatment (that the influent and effluent mean EMCs are statistically different from one another) and then to examine either a cumulative distribution function of influent and effluent quality or a standard parallel probability plot (Figure 4-1).

Before any efficiency plots are generated, appropriate nonparametric (or, if applicable, parametric) statistical tests should be conducted to determine whether any perceived differences in influent and effluent event mean concentrations are statistically significant (the level of significance should be provided, rather than noting only whether the result was significant; assume a 90% confidence level).

The effluent probability method is straightforward and provides a clear picture of the ultimate measure of BMP effectiveness—the effluent water quality. Curves of this type are the single most instructive piece of information that can result from a BMP evaluation study.

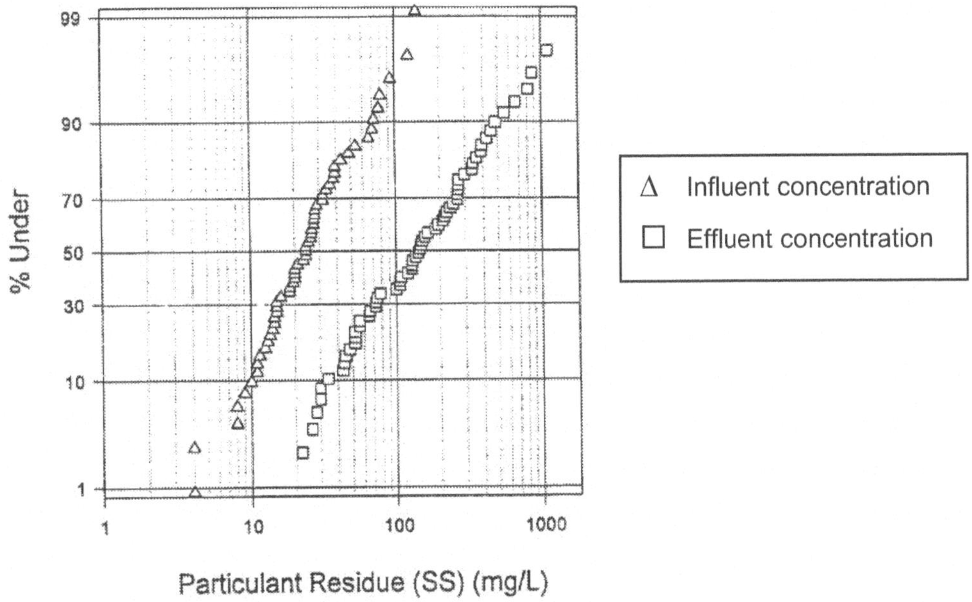

Figure 4-1. Probability Plot for Suspended Solids (ASCE, 2002).

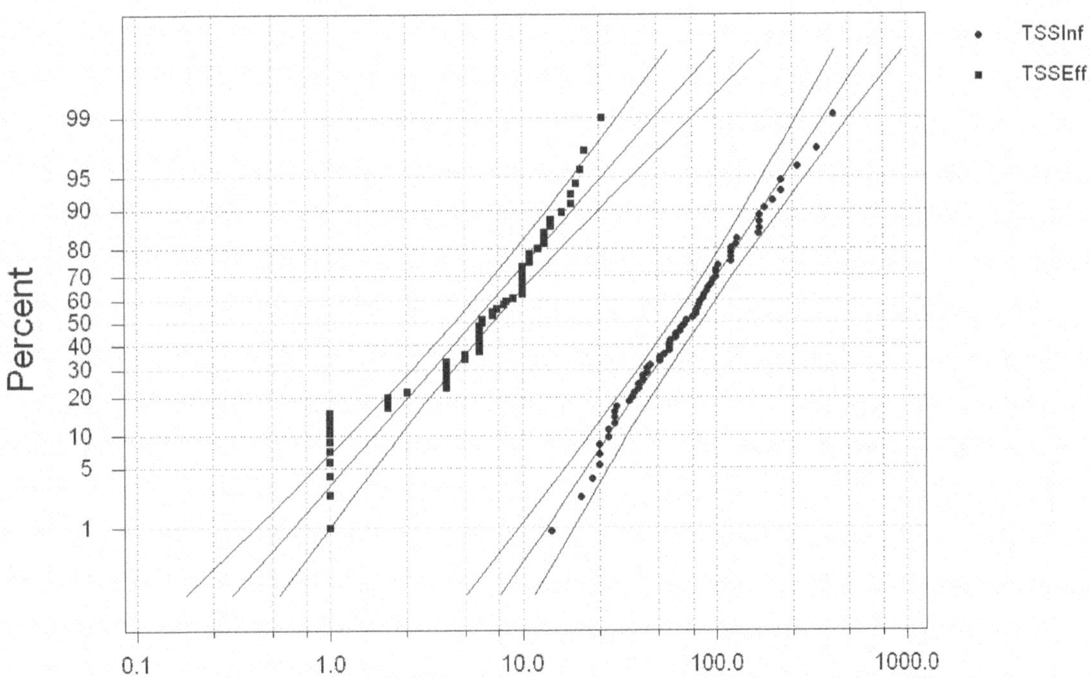

Figure 4-2. Probability Plot for Suspended Solids for Sand Filter (based on Caltrans data).

It is, however, necessary to be careful when interpreting probability plots. The probability plot of Figure 4-2 presents data for sand filters collected by Caltrans. There is a tendency to assume a correlation between the influent and effluent concentrations, so that the high influent concentrations result in high effluent concentrations. However, the plot of influent vs. effluent on Figure 4-3 shows that this is not the case. The effluent concentrations remain fairly constant over a range of influent concentrations.

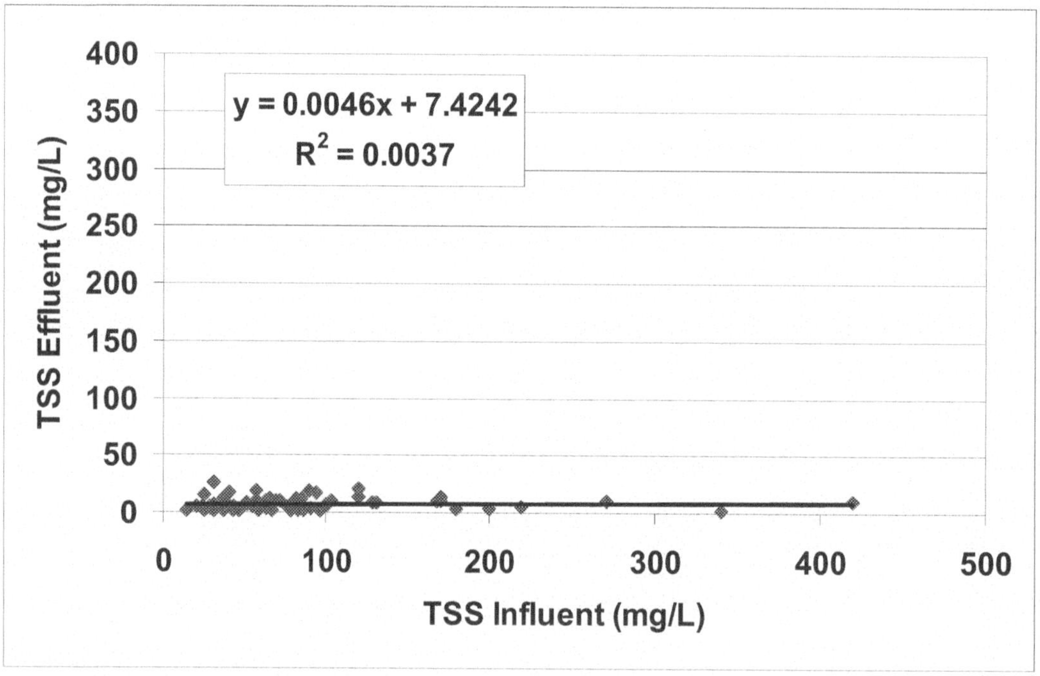

Figure 4-3. TSS Effluent vs. Influent for Sand Filter (based on Caltrans data).

4.4.6 Design of Monitoring System

Section 3.2 of ASCE manual "Developing a BMP Monitoring Plan," is included as an appendix and will be used as guidance when selecting monitoring locations.

4.4.7 Monitoring Frequency

The number of storms to be monitored each year, i.e., monitoring frequency, is an important consideration. In general, many measurements, i.e., many samples taken during many events, are needed to obtain enough data to be confident that actual BMP/SUDS performance is observed, rather than only noisy data, e.g., variability artifacts caused by external factors, equipment, and operator errors. Consequently, effectiveness studies can be expensive and time-consuming. Statistical confidence in the results of the monitoring program, attained by collecting samples from a significant number of events, should be assigned a higher importance than collecting information from a great number of locations or testing the water for a multitude of constituents.

4.4.8 Water Quality Sample Collection Techniques

4.4.8.1 Grab Samples

The term *grab sample* defines an individual sample collected within a short period of time at a particular location. Analysis of a grab sample provides a "snapshot" of stormwater quality at a single point in time. Grab samples are suitable for monitoring virtually all typical stormwater quality constituents. In fact, they are the only means of monitoring constituents that transform rapidly (requiring special preservation) or adhere to containers, such as oil and grease, total petroleum hydrocarbons (TPH), and bacteria.

- The results from a single grab sample generally are not sufficient to develop reliable estimates of the event mean pollutant concentration or pollutant load because stormwater quality tends to vary widely during a storm event. Nevertheless, grab sampling has an important role in many stormwater monitoring programs for the following reasons:

- A single grab sample collected during the early part of a storm can be used to characterize pollutants associated with the first flush. The highest pollutant concentrations are often contained in the runoff generated in the early part of a storm event, especially in small catchment areas with mostly impervious surfaces, and from storms with relatively constant rainfall. In such cases, the first flush may carry pollutants that have accumulated in the collection system and on paved surfaces during dry weather. Thus, the results from single grab samples of runoff collected at the beginning of the storm may be useful for identifying the pollutants that are present at concentrations of concern. This strategy may be less effective in areas of frequent low-intensity, long duration rainfall, where first flush effects are less obvious.

- Some measurable characteristics, such as temperature, pH, and the concentrations of total residual chlorine, phenols, volatile organic compounds (VOCs), and bacteria, transform or degrade so rapidly that compositing can introduce considerable bias. (Note: Grab sampling is the typical method for monitoring VOCs because VOCs can be lost through evaporation if the samples are exposed to air during compositing.)

- Some pollutants, such as oil and grease and TPH, tend to adhere to sample container surfaces; thus, transfer between sampling containers must be minimized; if program objectives require characterization of the average concentration of oil and grease over the duration of a storm, this information should be obtained from a series of grab samples analyzed individually.

To estimate event mean concentrations or pollutant loads, a series of grab samples could be collected at short time intervals throughout the course of a storm event. There are several different methods of obtaining information from a series of grab samples. One way would be to analyze each grab sample individually, to use the results to assess the rise and fall of pollutant concentrations during a storm, and to estimate event mean concentrations of pollutants. This method can be particularly useful if the monitoring objective is to discern peak pollutant concentrations or peak loading rates for assessing short-term water quality impacts. Because analyzing each grab sample separately adds significantly to laboratory costs, this method is rarely used except when detailed information about changes in constituent concentrations over the course of a storm is the objective.

4.4.8.2 Composite Samples

Another approach is to combine appropriate portions of each grab sample to form a single composite sample for analysis, but this is generally impractical if there are more than a few stations to monitor. Moreover, manual monitoring can be more costly than automated monitoring if the program encompasses more than a few storm events. For these reasons, many monitoring programs have found that the use of automated monitoring equipment and methods are more appropriate for compiling composite samples than manual monitoring. If detection of peak concentrations or loading rates is not essential, composite sampling can be more cost-effective for estimating event mean concentrations and pollutant loads. A composite sample is a mixture of several individual sample aliquots, collected at specific intervals of time or flow during a storm event and combined into a single sample. Thus, the composite sample integrates the effects of variations in stormwater quality that occur during a storm event. Compositing is suitable for typical stormwater quality constituents, except for those that transform rapidly, such as fecal coliforms, residual chlorine, pH, and volatile organic compounds, or those that adhere to container surfaces, such as oil and grease.

The two basic types of composite samples are time proportional and flow proportional. A time-proportional composite sample is prepared by collecting individual aliquots of equal volume at equal time intervals, e.g., every 20 minutes, during a storm event, and combining them into a single sample for laboratory analysis. Time-proportional samples do not take into account variations in flow; pollutant concentrations in the aliquots collected during the portion of the storm with lower flows are given the same weight as those collected during higher flows. Consequently, time proportional composite samples generally do not yield reliable estimates of event mean concentrations or pollutant loads, unless the interval between collecting the aliquots is very brief and flow rates are relatively constant.

Flow-weighted composite samples are better suited for estimating event mean concentrations and pollutant loads. A flow-weighted composite sample can be collected in several ways (U.S. EPA, 1992):

- Constant Time—Volume Proportional to Flow Rate—Aliquots are collected at equal increments of time during a storm event, and varying amounts of each aliquot are combined to form a single composite sample. The amount of water removed from each aliquot is proportional to the flow rate at the time the aliquot was collected. This type of composite sample can be collected using either manual or automated techniques.
- Constant Time—Volume Proportional to Flow Volume Increment—Aliquots are collected at equal increments of time during a storm event, and varying amounts of each aliquot are combined to form a single composite sample. The amount of water removed from each aliquot is proportional to the volume of flow since the preceding aliquot was collected. This type of compositing is generally used in conjunction with an automated monitoring system that includes a continuous flow measurement device. It can also be used for manual sampling in conjunction with a continuous flow measurement device, but this combination is uncommon.
- Constant Volume—Time Proportional to Flow Volume Increment—Aliquots of equal volume are taken at equal increments of flow volume (regardless of time) and

combined to form a single composite sample. This type of compositing is generally used in conjunction with an automated monitoring system that includes a continuous flow measurement device.

4.4.8.3 Automatic Sampling

Automated monitoring involves the use of electronic or mechanical devices that do not require the presence of an operator during sample collection. It is the preferred method for collecting flow-weighted composite samples. Automated samplers can be set to take a sample when a predetermined flow rate is reached. By setting the automated equipment to collect flow-weighted composite samples using the constant volume-time proportional to flow method, the need to measure samples for compositing can be reduced.

BMP/SUDS monitoring can be an especially useful application of equipment such as continuous flow recorders, automatic samplers, and continuous monitoring probes for the following reasons:

- Automated systems can generate data on virtually the entire volume of runoff that passes through the BMP/SUDS (i.e., they are not likely to omit small events and the beginnings and ends of other events).
- Automated systems are well suited for producing data sets that are useful (recognizing that performance evaluations are generally based on the differences between inlet and outlet concentration data sets, both of which are inherently noisy).
- The information generated through performance monitoring programs can be used to avoid inappropriate applications of BMP/SUDS, thereby justifying the costs associated with use of automated systems.

4.5 Qualitative Monitoring

4.5.1 Approach

Qualitative monitoring is a cost-effective method of studying BMP/SUDSs in time. The approach is not aimed at identifying the levels of specific pollutants at unique points within the system, but at evaluating the status of the system and the receiving watercourse. This evaluation can be carried on consistently within a wide range of sites and should provide important details on the following:

- The change in water quality of the BMP/SUDS with time;
- The change in hydraulic performance of the BMP/SUDS with time;
- The influence of maintenance on performance;
- The influence of particular design characteristics on performance;
- The impact of the system on receiving watercourse quality;
- The ecological status of the system;
- The aesthetic benefits of the systems;
- Any risks associated with the systems; and
- Appropriate operation and maintenance schedules.

4.5.2 Qualitative Observations

Because of differences in the design features, physical and chemical pollutant removal processes, and certain site-specific factors, it will be necessary to record different qualitative observations at each monitored BMP/SUDS locations. Some of the questions that can be answered based on empirical observation of selected BMPs are listed below. Based on the recommendations for Phase 2 monitoring, similar questions may be developed for other types of treatment systems.

4.5.2.1. Extended Detention Basins

- Does the basin drain from a full condition within the design draining time?
- Does it drain completely, with no standing water remaining?
- How well do various basin outlet configurations perform, and how well do they resist clogging?
- Do energy dissipators prevent resuspension of material near the basin inlet?
- Is it possible to observe whether the basin geometry or other factors reduce/eliminate short-circuiting?
- How does the basin perform in the winter: Does ice accumulate or obstruct the basin outlet?
- What is the required maintenance frequency?

To answer these questions, at least the following information would be recorded:

- Depth of water in the basin;
- Time since last storm event;
- Total rainfall from last storm event;
- Time to completely drain a full basin;
- Persistence of small pools or other saturated areas several days after a rainfall event;
- Height and density of vegetation at the basin invert;
- Evidence of outlet clogging;
- Locations and amounts of accumulated sediment;
- Evidence of resuspension of sediment near basin inlet;
- Types and amounts of trash and debris removed during maintenance; and
- Evidence of erosion, scour, or slope instability.

4.5.2.2 Sand Filters

- Does the facility drain fully within the design draining time?
- How do the filter sizing criteria affect the maintenance interval?
- Can variations in clogging be related to filter media grain size?
- How does the basin perform in the winter: Does ice accumulate, obstruct the outlet, or interfere with proper operation?

Relevant observations would include the following:

- Depth of water in the basin;
- Time since last storm event;
- Total rainfall from last storm event;

- Time to completely drain a full basin;
- Persistence of small pools or other saturated areas several days after a rainfall event;
- Evidence of solids accumulation on the filter surface;
- Depth of penetration of solids into the filter bed;
- Height and density of vegetation at the basin invert;
- Evidence of filter clogging;
- Locations and amounts of accumulated sediment;
- Evidence of sediment resuspension near basin inlet;
- Types and amounts of trash and debris removed during maintenance; and
- Evidence of erosion, scour, or slope instability.

4.5.2.3 Infiltration Devices

- Does the device drain completely within the design draining time when new?
- How does draining time change as the facility ages?
- Can such change be related to soil/subsurface properties?
- How does the device perform in the winter: Does ice accumulate in the device or obstruct the outlet?
- Does vegetation seem to help maintain initial infiltration rates?

Relevant observations would include the following:

- Depth of water in the basin;
- Time since last storm event;
- Total rainfall from last storm event;
- Time to completely drain a full basin;
- Persistence of small pools or other saturated areas several days after a rainfall event;
- Characteristics of the soil;
- Depth to groundwater below the basin invert;
- Height and density of vegetation at the basin invert;
- Evidence of basin clogging;
- Locations and amounts of accumulated sediment;
- Types and amounts of trash and debris removed during maintenance; and
- Evidence of erosion, scour, or slope instability.

4.5.2.4 Wet Ponds

- Does vegetation establish rapidly and remain confined to designated areas such as the perimeter bench?
- Is there evidence that extensive vegetated cover promotes short circuiting of flow?
- How does it perform in the winter: Does ice accumulate in the basin or obstruct the outlet?
- Is removal of vegetation necessary for performance or control of nutrients?

Relevant observations would include the following:

- Depth of water in the pond;
- Time since last storm event;

- Total rainfall of last storm event;
- Time to drain a full pond;
- Height and density of vegetation on the pond invert;
- Distribution of emergent vegetation;
- Evidence of short circuiting or dead areas caused by vegetation;
- Evidence of outlet clogging;
- Locations and amounts of accumulated sediment;
- Evidence of resuspension of sediment near inlet;
- Types and amounts of trash and debris removed during maintenance; and
- Evidence of erosion, scour, or slope instability.

4.5.2.5 Swales

- Are current guidelines of slope and width conducive to a stable cross-section without channelization?
- Are differences in stability related to mowing frequency or types of vegetation?
- How does the swale perform in the winter: Does ice accumulate in the swale or obstruct its outlet?
- Are nutrients exported during dormancy of vegetation?

Relevant observations would include the following:

- Time since last storm event;
- Total rainfall from last storm event;
- Height and density of vegetation in the swale;
- Date of last mowing;
- Evidence of channelization;
- Locations and amounts of accumulated sediment;
- Evidence of sediment resuspension near inlet;
- Types and amounts of trash and debris removed during maintenance; and
- Evidence of erosion, scour, or slope instability.

4.5.2.6 Buffer Strips

- Are current guidelines of slope and width conclusive to stability?
- Are differences in stability related to mowing frequency or types of vegetation?
- How do the strips perform in the winter: Does ice accumulate in the basin or obstruct the outlet?
- Does export of nutrients occur during vegetation dormancy periods?

Relevant observations would include the following:

- Time since last storm event;
- Total rainfall from last storm event;
- Height and density of vegetation;
- Date of last mowing;
- Types and amounts of trash and debris removed during maintenance; and
- Evidence of erosion, scour, and slope instability.

4.5.2.7 Porous Pavement

- How does infiltration rate change over time?
- What effect does periodic cleaning have on infiltration rate?
- How does the pavement hold up to various volumes and loads of traffic (trucks)?

Relevant observations would include the following:

- Time since last storm event;
- Total rainfall from last storm event;
- Discharge rate from underdrain;
- Evidence of instability attributable to traffic loadings;
- Evidence of reduced infiltration rates (more surface runoff); and
- Time since cleaning or rejuvenation of surface.

4.5.3 Observation Collation Sheets

To date, qualitative approaches to BMP/SUDS monitoring have been implemented in both the United States and Scotland. The Scottish data collation sheets have been expanded and improved for use across the United Kingdom to allow consistent compilation and interpretation of a large amount of data. A slightly different approach has been used in California, which may also benefit this study. The final data collation sheets will be developed during the Phase 2 study and are likely to be an amalgamation of the two approaches to ensure consistency between the United Kingdom and the United States. It will be important to glean information both on a periodic basis and during specific storm events.

The following are examples of some of the types of forms that will be developed for use in the Phase 2 project.

FORM C

Caltrans BMP Retrofit Pilot Program

BMP Site Inspection Checklist for Extended Detention Basins

GENERAL INFORMATION:		
Date:	Time In:	Time Out:
Team Leader's Initials:	Stormwater Consultant:	
BMP Location:		

SEDIMENT/EROSION CONTROL:			Maintenance Required?
Sediment depth (is average sediment depth in excess of 18 inches, 10% of basin volume or is there an alluvial type deposit impeding EDB hydraulics at the inlet):			☐ yes ☐ no
Sediment type and location:			
Standing water or other drainage problems? Comments:	☐ yes	☐ no	☐ yes ☐ no
If yes, what is the depth of water? What is the extent of the water? Where is the location of the water?			
Evidence of erosion, rilling, or undercutting?	☐ yes	☐ no	☐ yes ☐ no
Seeps?	☐ yes	☐ no	☐ yes ☐ no
Embankment slumping or cracking?	☐ yes	☐ no	☐ yes ☐ no
Comments:			

VEGETATION:			Maintenance Required?	
Height of vegetation *(if length >18 inches mark yes):*			☐ yes	☐ no
Does vegetation cover all of the detention basin's floor and/or slopes? Is vegetation management required?	☐ yes	☐ no	☐ yes	☐ no
Is native vegetation, weeds, leaves, or other organic debris present in the basin? If yes, what type:	☐ yes	☐ no	☐ yes	☐ no
Does native vegetation, weeds, leaves, or other organic debris interfere with the operation of the detention basin by (a) overgrowing the detention basin vegetation, (b) interfering with the hydraulic residence time in the detention basin, or (c) saturation of the bottom with dead or dying vegetation, or (d) in another way? If so, explain:	☐ yes	☐ no	☐ yes	☐ no
Evidence of flattening of vegetation? If so, which species?	☐ yes	☐ no	☐ yes	☐ no
Condition of landscaping? ☐ Inhibited Growth ☐ Excessive Growth ☐ Normal Growth Comments:			☐ yes	☐ no

FORM C

Caltrans BMP Retrofit Pilot Program

BMP Site Inspection Checklist for Extended Detention Basins (continued)

STRUCTURAL:	Maintenance Required?
Are flumes, flume-to-basin transitions, basin -to-flume transitions, and energy dissipators clear of sediment? ☐ yes ☐ no	☐ yes ☐ no
Risers and Conveyances: Blockage or sediment in risers?: ☐ yes ☐ no If yes, extent and location: ________	☐ yes ☐ no
Is corrosion present? ☐ yes ☐ no ☐ not applicable If yes, extent and location: ________	☐ yes ☐ no
Evidence of concrete scour, spalling, or cracking of structural parts? ☐ yes ☐ no	☐ yes ☐ no
Fences: ☐ OK ☐ not OK ☐ not applicable	☐ yes ☐ no
Locks: ☐ OK ☐ not OK ☐ not applicable	☐ yes ☐ no
Other general maintenance concerns/comments:	

ANIMAL CONCERNS:	Maintenance Required?
Is there evidence of small animals (burrows, droppings, trails, gnawing marks, or stained rub marks)? ☐ yes ☐ no	☐ yes ☐ no
Comments:	

AESTHETIC CONCERNS:		
Debris (non-trash)?	☐ yes ☐ no	Type, amount and location:
Trash?	☐ yes ☐ no	Type, amount and location:
Graffiti?	☐ yes ☐ no	Description and location:
Other general aesthetic concerns:		

Water Management:

☐ ***Dewatering (describe):***

Start time:	End time:	Total time:	More work required? ☐ yes ☐ no
Equipment used:			

PHOTOGRAPHS (mandatory if maintenance is required):

Date	**Time**	**Description**	**Direction Facing**	**Roll**	**Photograph No.**

FORM C

Caltrans BMP Retrofit Pilot Program

BMP Site Inspection Checklist for Extended Detention Basins (continued)

Comments:

Condition of facility: ☐ acceptable ☐ not acceptable

Comments:

Dates for maintenance to be completed by:

(Team Leader's Signature)

FORM H

Caltrans BMP Retrofit Pilot Program

Empirical Observations Field Data Log Sheet for Storm Events for Extended Detention Basins

GENERAL INFORMATION:

Date:	Time in:
Team leader:	Stormwater Consultant:
Other personnel:	Location of facility:

METEOROLOGICAL CHARACTERISTICS:

Present rainfall: ☐ None ☐ Mist ☐ Drizzle ☐ Sprinkle ☐ Light, Steady ☐ Moderate ☐ Heavy

Time present rain event started: __________

Wind speed: ☐ 0-5 mph (light) ☐ 5-10 mph (moderate) ☐ 10-15 mph (brisk) ☐ >15 mph (strong)

Cloud cover: % Temperature ______________

Distribution of Rainfall: ☐ Uniform ☐ Localized cells ☐ Scattered showers

Comments:

HYDROLOGIC AND HYDRAULIC CHARACTERISTICS:

Flow conditions (*check all that apply*):

☐ Runoff entering facility

☐ Facility discharging through regular outlet

☐ Facility discharging via emergency overflow Time occurred: _______________

☐ Flow bypassing facility Time occurred: _______________

Standing water conditions (*record measurements as appropriate and check all that apply*):

☐ Water standing over entire basin. Depth: ______________________

☐ Other standing water pool(s): Describe Location: ______________________

Depth: ______________________

If during a Storm Event:

Record maximum depth of pooled water: Depth: ______________________

Comments: ______________________

____

INLET CONDITIONS:

Describe any erosion or resuspension of settled solids caused by flow in the influent channel and/or the basin:

____

Describe any obstructions or restrictions interfering with inflow/influent: ______________________

____

INFLUENT WATER QUALITY APPEARANCE:

(Check all that apply and describe under comments)

Odor:

☐ Hydrogen sulfide ☐ Musty ☐ Sewage ☐ Ammonia ☐ Hydrocarbons ☐ Pesticide or herbicide ☐ None

FORM H

Caltrans BMP Retrofit Pilot Program
Empirical Observations Field Data Log Sheet for Storm Events for Extended Detention Basins

Floating materials:
☐ Oil and grease (see below) ☐ Trash or debris ☐ Surface film ☐ Organic material ☐ None

Oil and grease:
☐ Sheen ☐ Heavy floating concentration ☐ Emulsion ☐ Deposit ☐ None

Color:
☐ Colorless ☐ Red ☐ Orange ☐ Yellow ☐ Green ☐ Blue ☐ Violet ☐ Brown ☐ Black
☐ Gray ☐ White ☐ Other: __________

Turbidity
☐ None ☐ Some cloudiness but transparent ☐ Cloudy, translucent ☐ Heavy cloudiness, opaque

Comments:

SOLIDS DEPOSITION AND RESUSPENSION:

For the following locations, record the type (trash or debris, oil and grease, other organics), location(s), area(s) covered, and depth(s), as applicable:

In inflow/influent channel: ______________________________

____

Over entire basin: ______________________________

____

In one spot: ______________________________

____

In multiple spots: ______________________________

____

In outflow/effluent channel: ______________________________

____

Solids resuspension evident (*check all that apply*):
☐ In inflow/influent channel ☐ Near inlet ☐ Near outlet ☐ In outflow/effluent channel
☐ Other (describe): ______________________________

Comments:

FORM H
Caltrans BMP Retrofit Pilot Program
Empirical Observations Field Data Log Sheet for Storm Events for Extended Detention Basins

EROSION:

For the following locations, record erosion locations, area(s) covered, and depth(s), as applicable:

In inflow/influent channel: ______________________________

____

Near inlet: ______________________________

____

Basin bed: ______________________________

____

Basin side slopes: ______________________________

____

Near outlet: ______________________________

____

In outflow/effluent channel: ______________________________

Comments: ______________________________

____

VEGETATION CONDITION:

Basin bed vegetation cover:

☐ Complete ☐ Few small bare spots ☐ Few large bare spots ☐ Many small bare spots
☐ Many large bare spots ☐ Large areas bare ☐ All or nearly bare

Basin side slopes vegetation cover:

☐ Complete ☐ Few small bare spots ☐ Few large bare spots ☐ Many small bare spots
☐ Many large bare spots ☐ Large areas bare ☐ All or nearly bare

Basin bed vegetation type:

☐ All grasses ☐ Mostly grasses, some wetland plants ☐ Mostly wetland plants, some grasses
☐ All wetland plants

Extent of woody shrubs or trees: ______________________________

Basin side slopes vegetation type:

☐ All grasses ☐ Mostly grasses, some wetland plants ☐ Mostly wetland plants, some grasses
☐ All wetland plants

Extent of woody shrubs or trees: ______________________________

Comments:

OUTLET CONDITIONS:

Describe any obstructions or restrictions interfering with outflow/effluent:

FORM H

Caltrans BMP Retrofit Pilot Program
Empirical Observations Field Data Log Sheet for Storm Events for Extended Detention Basins

EFFLUENT WATER QUALITY APPEARANCE:
Check all that apply and describe under comments

Odor:
☐ Hydrogen sulfide ☐ Musty ☐ Sewage ☐ Ammonia ☐ Hydrocarbon ☐ Pesticide or herbicide ☐ None

Floating materials:
☐ Oil and grease (see below) ☐ Trash or debris ☐ Surface film ☐ Organic material ☐ None

Oil and grease:
☐ Sheen ☐ Heavy floating concentration ☐ Emulsion ☐ Deposit ☐ None

Color:
☐ Colorless ☐ Red ☐ Orange ☐ Yellow ☐ Green ☐ Blue ☐ Violet ☐ Brown ☐ Black
☐ Gray ☐ White ☐ Other: __________

Turbidity
☐ None ☐ Some cloudiness but transparent ☐ Cloudy, translucent ☐ Heavy cloudiness, opaque

Water quality appearance comments:

VECTORS:

Presence of vectors: ☐ Yes ☐ No
If yes, what type: ☐ Mosquitoes ☐ Blackfly ☐ Cockroaches ☐ Rats
☐ Other______________________________

Comments:

STRUCTURAL CONDITION OF FACILITY:

Record the presence of the following (*check all that apply and give location in comments*):
☐ Inlet structure damage ☐ Outlet structure damage ☐ Vandalism
Comments:

MONITORING EQUIPMENT CONDITIONS:

Equipment functional: ☐ Yes ☐ No ☐ N/A
Comments:

CHANGES DURING STORM SINCE LAST OBSERVATION:

Date:______________________ Time: ______________________
Comments:

Date:______________________ Time: ______________________
Comments:

FORM H

Caltrans BMP Retrofit Pilot Program
Empirical Observations Field Data Log Sheet for Storm Events
for Extended Detention Basins

ADDITIONAL COMMENTS:

____
____
____
____
____
____
____
____

____
(Team Leader's Signature)

Caltrans CHD Biofilter Strip Evaluation Program
Site Inspection Field Data Log Sheet for Biofiltration Strips (Quarterly)

Date:	Time in:	Time out:

VEGETATION CONDITION:

Strip bed vegetation cover:

☐ Complete ☐ Few small bare spots ☐ Few large bare spots ☐ Many small bare spots

☐ Many large bare spots ☐ Large areas bare ☐ All or nearly bare

Strip bed vegetation type:

☐ All grasses ☐ Mostly grasses, some wetland plants ☐ Mostly wetland plants, some grasses

☐ All wetland plants

Extent of woody shrubs or trees: ______________________________

Strip vegetation height: ___________ inches

% Strip vegetation dormant (brown) ______%

Comments:

SOLIDS DEPOSITION AND RESUSPENSION:

For the following locations, record the type (trash or debris, oil and grease, other organics), location(s), area(s) covered, and depth(s), as applicable:

In collection ditch: ______________________________

____

Over entire strip: ______________________________

____

In one spot: ______________________________

____

In multiple spots: ______________________________

____

Solids resuspension evident (check all that apply):

☐ In inflow/influent channel ☐ Near inlet ☐ Near outlet ☐ In outflow/effluent channel

☐ Other (describe): ______________________________

Comments:

Site:____________________System:____________

Caltrans CHD Biofilter Strip Evaluation Program
Site Inspection Field Data Log Sheet
for Biofiltration Strips (Quarterly) (continued)

Date:	Time in:	Time out:

EROSION:

For the following locations, record erosion locations, area(s) covered, and depth(s), as applicable:

Fire break: ______________________________

Near inlet: ______________________________

Strip bed: ______________________________

At strip/effluent channel interface : ______________________________

Comments:

STRUCTURAL CONDITION OF FACILITY:

Record the presence of the following (check all that apply and give location in comments):

☐ Collection ditch damage ☐ Vehicle damage ☐ Vandalism

Comments:

Site:______________________System:____________

Caltrans CHD Biofilter Strip Evaluation Program
Empirical Observations Field Data Log Sheet for Storm Events for Biofiltration Strips

GENERAL INFORMATION:

Date:	Time in:	Time out:
Team leader's initial:	Stormwater consultant:	
Other personnel:	Temperature:	

Purpose of visit (check all that apply): ☐ Storm monitoring ☐ Routine inspection ☐ Special inspection ☐ Maintenance

HYDROLOGIC AND HYDRAULIC CHARACTERISTICS:

Flow conditions (check all that apply):
☐ Runoff entering facility
☐ Flow distributed uniformly throughout Strip
☐ Flow not uniformly distributed throughout Strip

If flow is not uniformly distributed throughout the Strip, describe distribution (e.g., where concentrated, where channelized, approximate dimensions and depths of concentrations and/or channel):

Flow conditions continued (check all that apply):
☐ Water flowing through Strip at a perceptible velocity
☐ Facility discharging
☐ Flow bypassing facility
☐ Flow overtopping facility

Comments:

Standing water conditions (record measurements as appropriate):
☐ Water standing over entire Strip. Depth: __________
☐ Water standing in one isolated pool. Describe Location: ______________________ Depth: __________
☐ Water standing in multiple pools: Describe Locations: ______________________ ______________________ Depths: ______________________

Comments:

Photos Taken ☐ (During post-storm visit)

Problems:

____
(Team Leader's Signature)

SUDS/BMPs Site Visit Sheets: Type 1
(For Wetlands, Ponds, Swales, Detention & Infiltration Basins

Site : ______________ SUD Ref: ____________ Date: ________________ Time:______________

Weather Conditions: ________________ Weather Prior to Visit: ____________________

Air Temperature: ________________ Water temperature: ________________

	1	2	3	4	5	DESCRIPTION
INLET(S) / INFLOW						
1. Inflow rate						
2. Visible Pollutants: *a. Oil film on surface* *b. Oils trapped in vegetation* *c. Smell of oils* *d. Construction materials (e.g., cement, topsoil)* *e. Dyes/paints/tars* *f. Detergents* *g. Other:*						
3. Pollution Indicators: *a. Algae (long filaments)* *b. Algae (unicellular)* *c. Blue/green algae* *d. Secchi depth / turbidity* *e. Gas bubbling to surface* *f. Smell (e.g., rotten eggs)* *g. Other:*						
4. Sediment accumulation						
5. Floating vegetation/grass cuttings						
6. Erosion						
7. Litter (describe)						
8. Waste (e.g., sink, sanitary towels, condoms)						
9. Clogging of inlet						
OUTLET(S) / OUTFLOW						
1. Outflow rate						

SUDS/BMPs Site Visit Sheets: Type 1 (cont.)
(For Wetlands, Ponds, Swales, Detention & Infiltration Basins

2. Visible Pollutants: *a.* Oil film on surface *b.* Oils trapped in vegetation *c.* Smell of oils *d.* Construction materials (cement ,topsoil) *e.* Dyes/paints/tars *f.* Detergents *g.* Other:						
3. Pollution Indicators: *a.* Algae (long filaments) *b.* Algae (unicellular) *c.* Blue/green algae *d.* Secchi depth/turbidity *e.* Gas bubbling to surface *f.* Smell (e.g., rotten eggs) *g.* Other:						
4. Sediment accumulation						
5. Floating vegetation/grass cuttings						
6. Erosion						
7. Litter (describe)						
8. Waste (e.g., sink, sanitary towels, condoms)						
9. Clogging of outlet						
GENERAL SUD/BMP CONDITION	**1**	**2**	**3**	**4**	**5**	**DESCRIPTION**
1. Water color						
2. Water level/soil moisture content						
3. Litter buildup						
4. Vandalism						
5. Erosion (describe)						
6. Pollutants (describe)						

SUDS/BMPs Site Visit Sheets: Type 1 (cont.)
(For Wetlands, Ponds, Swales, Detention & Infiltration Basins

7. Pollution indicators (describe)						
8. Sediment accumulation						
9. Odor						
10. Health of vegetation						
11. Volume of vegetation						
12. Evidence of amenity activities						

SUD/BMP ECOLOGY	**DESCRIPTION**
Mayflies	
Damselflies/dragonflies	
Beetles	
Mosquitoes/gnats (describe where)	
Other insects/mammals	
Birds:	
Reeds	
Bullrushes	
Waterlilies	
Other plants:	
HYDRAULIC PERFORMANCE	**DESCRIPTION**
Max water level since previous visit	
Evidence of overtopping	
Evidence of bypassing	
Extent of any reported flooding	
MAINTENANCE	**From Observation or Interviews with Site Managers**
Grass cutting (for appearances)	
Grass cutting (for performance)	
Vegetation control (not grass)	
Aquatic plant management	
Litter collection (number of bags?)	
Debris removal	
Rodding/unblocking inlets/outlets	

SUDS/BMPs Site Visit Sheets: Type 1 (cont.)
(For Wetlands, Ponds, Swales, Detention & Infiltration Basins

Silt removal (from infrastructure)	
Silt removal (from SUDS)	
Algae/weed removal	
Erosion/scour repairs	
Scarifying of surfaces	
Fence/sign maintenance	
Irregular rehabilitation works (describe)	
Other: (describe)	

CATEGORY	**1**	**2**	**3**	**4**	**5**	**DESCRIPTION**
RECEIVING WATERCOURSE						
1. Flow rate						
2. Visible Pollutants: *a.* Oil film on surface *b.* Oils trapped in vegetation *c.* Smell of oils *d.* Construction materials (e.g., cement, topsoil) *e.* Dyes/paints/tars *f.* Detergents *g.* Other:						
3. Pollution Indicators: *a.* Algae (long filaments) *b.* Algae (unicellular) *c.* Blue/green algae *d.* Secchi depth/turbidity *e.* Gas bubbling to surface *f.* Smell (e.g., rotten eggs) *g.* Other:						
4. Sediment accumulation						
6. Erosion						

RECEIVING WATERCOURSE ECOLOGY	
Mayflies	
Damselflies/dragonflies	
Beetles	

SUDS/BMPs Site Visit Sheets: Type 1 (cont.)
(For Wetlands, Ponds, Swales, Detention & Infiltration Basins

Mosquitos/gnats (describe where)	
Other insects/mammals:	
Birds:	
Reeds	
Bullrushes	
Waterlilies	
Other plants:	
Comments:	

SUDS/BMPs Site Visit Sheets: Type 2
(For Filter Drains, Infiltration Trenches, Soakaways, Permeable Pavements)

Site : ______________ **SUD Ref:** ____________ **Date:** ________________ **Time:**______________

Weather Conditions: ________________ **Weather Prior to Visit:** ________________

Air Temperature: ________________ **Water Temperature:** ________________

	1	2	3	4	5	DESCRIPTION
INLET(S) / INFLOW (if > 1, give general description, or numbers in each category)						
Inflow rate						
2. Visible Pollutants: *a. Oil film on surface* *b. Oils trapped in vegetation* *c. Smell of oils* *d. Construction materials (e.g., cement, topsoil)* *e. Dyes/paints tars* *f. Detergents* *g. Other:*						
Sediment accumulation						
4. Vegetation/grass cuttings						
5. Erosion						
6. Litter (describe)						
7. Waste (e.g., sink, sanitary towels, condoms)						
8. Clogging of inlet						
OUTLET(S) / OUTFLOW (if > 1, give general description, or numbers in each category)						
Outflow rate						

SUDS/BMPs Site Visit Sheets: Type 2 (cont.)
(For Filter Drains, Infiltration Trenches, Soakaways, Permeable Pavements)

2. Visible Pollutants: *a. Oil film on surface* *b. Oils trapped in vegetation* *c. Smell of oils* *d. Construction materials (cement ,topsoil)* *e. Dyes/paints/tars* *f. Detergents* *g. Other:*						
3. Sediment accumulation						
4. Floating vegetation/grass cuttings						
5. Erosion						
6. Litter (describe)						
7. Waste (e.g., sink, sanitary towels, condoms)						
8. Clogging of outlet						
SILT TRAPS (if > 1, give general description, or numbers in each category)						**DESCRIPTION**
1. Flow rate						
2. Visible Pollutants: *a. Oil film on surface* *b. Oils trapped in silts* *c. Smell of oils* *d. Construction materials (cement, topsoil)* *e. Dyes/paints/tars* *f. Detergents* *g. Other:*						
3. Sediment/silt accumulation						
4. Floating vegetation/grass cuttings						
5. Erosion						
6. Litter (describe)						
7. Waste (e.g., sink, sanitary towels, condoms)						
8. Clogging of inlet/outlet						
GENERAL SUD/BMP CONDITION	**1**	**2**	**3**	**4**	**5**	**DESCRIPTION**
1. Surface settlement/damage						
2. Water level/moisture content of fill						
3. Litter buildup						

SUDS/BMPs Site Visit Sheets: Type 2 (cont.)
(For Filter Drains, Infiltration Trenches, Soakaways, Permeable Pavements)

4. Vandalism						
5. Erosion (describe)						
6. Visible pollutants (describe)						
8. Surface sediment accumulation						
9. Odor						
HYDRAULIC PERFORMANCE	**DESCRIPTION**					
Max water level since previous visit						
Any evidence of overtopping						
Any evidence of bypassing structure						
Extent of any reported flooding						
MAINTENANCE	**From Observation or via Interviews with Site Managers**					
Grass cutting (for amenity)						
Grass cutting (for performance)						
Street sweeping						
Gully pot cleaning						
Litter collection (number of bags?)						
Debris removal						
Rodding/unblocking inlets/outlets						
Silt removal (from infrastructure)						
Silt removal (from SUDS)						
Algae/weed removal						
Erosion/scour repairs						
Fence/sign maintenance						
Irregular rehabilitation (describe):						
Other (describe):						
CATEGORY	**1**	**2**	**3**	**4**	**5**	**DESCRIPTION**
RECEIVING WATERCOURSE						
1. Flow rate						

SUDS/BMPs Site Visit Sheets: Type 2 (cont.)
(For Filter Drains, Infiltration Trenches, Soakaways, Permeable Pavements)

2. Visible Pollutants: *a. Oil film on surface* *b. Oils trapped in vegetation* *c. Smell of oils* *d. Construction materials (e.g., cement, topsoil)* *e. Dyes/paints/tars* *f. Detergents* *g. Other:*						
3. Pollution Indicators: *a. Algae (long filaments)* *b. Algae (unicellular)* *c. Blue/green algae* *d. Secchi depth/turbidity* *e. Gas bubbling to surface* *f. Smell (e.g., rotten eggs)* *g. Other:*						
4. Sediment accumulation						
5. Erosion						

RECEIVING WATERCOURSE ECOLOGY	
Mayflies	
Damselflies/dragonflies	
Beetles	
Mosquitoes/gnats (describe where)	
Other insects/mammals:	
Birds:	
Reeds	
Bullrushes	
Waterlilies	
Other plants:	

Comments:

SUDS Qualitative Observations: Explanatory Notes (Sheet 1 of 5)

Category	Description	1	2	3	4	5
INLET / OUTLET						
1. Flow rate	Rate of pipe flow	No flow	Trickle	Quarter depth	Half depth	Full bore
2. Visible pollutants	Pollutants that are visible at point of inflow	No evidence of pollution	Individual small deposit/very faint smell	2/3 individual small deposits/mild smell	Many deposits, some larger, strong smell	Mass of pollution/ very strong odor
3. Pollution indicators	Evidence of SUDS response to pollution	No algae/clear waters/no gas/no odor	0-5% of water surface area/some cloudy patches/occasional bubble/very faint odor	5-10% of water surface area/some turbidity/a few bubbles/mild odor	10-25% of water surface area/significant turbidity/some gas/strong odor	25%-50% of water surface area/high turbidity/continuous bubbles/very strong odor
4. Sediment accumulation	Build-up of silts/sediments just below inlet	No evidence of silts/sediments	Evident at base of structure	Covering base of structure	Up to pipe invert level	Beginning to block piping
5. Floating vegetation	Floating vegetative matter e.g., grass cuttings	No floating vegetation	Single occurrence	<5 occurrences	Affecting 25% of SUDS surface	Affecting > 50% of SUDS surface
6. Erosion	Extent of damage caused to SUD structure (banks, base, etc.) by flows or rainfall	No erosion	Initial evidence of erosion	Increasing levels of erosion	Erosion damage, which in the future is likely to impair SUDS performance (describe)	Significant erosion damage, impairing SUDS performance (describe)
7. Litter	Litter type (indicate whether construction debris or wind-blown)	No litter	Single piece	2-3 pieces	Some influence on SUDS performance (say how many pieces)	Significant influence on SUDS performance (number of pieces)
8. Waste	Extent of debris accumulation (indication of wrong connections)	No solid waste debris	Single piece	2-3 pieces	Some influence on SUDS performance (say how many pieces)	Significant influence on SUDS performance (number of pieces)

SUDS Qualitative Observations: Explanatory Notes (Sheet 2 of 5)

9. Clogging of outlet structure	Extent to which inlet/outlet is blocked by e.g., litter, vegetation, plastic, silts, etc	No clogging	Some hindrance to flow	25% of flow path clogged	50% of flow path clogged	Completely clogged
GENERAL CONDITION	**Description**	**1**	**2**	**3**	**4**	**5**
1. Water color	Color of water	Clear	Brown	Grey	Pale Green	Dark Green
2. Water level/soil moisture	Water level/wetness in base of swale	Very low/parched	< pipe levels (say how far)/dry base	At pipe level/wet base	Above outlet pipe level/flowing	Overtopping/full of water
3. Litter buildup	Extent of litter buildup across swale	No litter	Individual piece	2-3 pieces	Some influence on SUDS performance (number of pieces)	Significant influence on SUDS performance
4. Vandalism	Damage to vegetation, sign boards, fencing, fires, other (specify)	No vandalism	Very minor damage	Some noticeable damage	Significant damage: influence on site aesthetics	Major influence on aesthetics and performance
5. Erosion	Extent of damage caused to SUD structure (banks, base, etc) by flows or rainfall	No erosion	Initial evidence of erosion	Increasing levels of erosion	Erosion damage, which in the future is likely to impair SUDS performance (describe)	Significant erosion damage, impairing SUDS performance
6. Pollutants	Pollutants that are visible within SUD	None	Single small deposits	Several small deposits	Many deposits, some large	Mass of pollution
7. Pollution indicators	Evidence of SUD response to pollution	No algae/clear waters	0-5% of water surface area/ occasional bubble /very faint odor	5-10% of water surface area/a few bubbles/mild odor	10-25% of water surface area/some gas/strong odor	25-50% of water surface area/continuous bubbles/very strong smell
8. Sediment accumulation	Visible sediments in areas away from inlets/outlets	None	Individual small deposit	Several small deposits	Several significant larger deposits	Significant accumulation over DWF/base of structure

SUDS Qualitative Observations: Explanatory Notes (Sheet 3 of 5)

9. Odor	Rotten eggs or oils	None	Very faint odor	Mild odor	Strong odor	Very strong smell
10. Health of vegetation	Health of vegetation in SUD (with reasons)	Very poor	Poor	Satisfactory	Good	Very good
11. Volume of vegetation	Density of vegetation colonizing SUD (specify types)	None	Occasional marginal plants	25% coverage	50% coverage	Complete coverage. Significant % of capacity
12. Evidence of amenity activities	Extent site is visited by locals (who, reasons)	None	Infrequently	Frequently by a limited number	Frequently by several	Constantly, by many
MAINTENANCE	**Description**					
All items	Describe: Details of maintenance activities undertaken since previous visit. Number of times such activities have been carried out. Success of activities. Any positive or negative impact on SUDS.					
ECOLOGY	**Description**					
All items	Please describe: How many of each species were viewed/density of extent of particular plants. If the species were viewed in a particular place. If there was a specific feature of the SUDS that seemed to be encouraging/promoting their growth.					
HYDRAULIC PERFORMANCE	**Description**					
Max water level	This should be determined from water marks/trash marks/erosion marks, or from discussion with locals/site managers.					
Any evidence of overtopping	Any evidence of where the capacity of the SUDS structure has been exceeded, and bypassing or overflow mechanisms have been utilized. Discussion with locals/site managers may also provide relevant information. Description of any damage caused by the overtopping.					

SUDS Qualitative Observations: Explanatory Notes (Sheet 4 of 5)

Extent of any reported flooding	Any evidence of where, due to extreme rainfall, the SUD design has been overwhelmed, and flooding has occurred Extent of the flooding Damage caused by the flooding					
RECEIVING WATERCOURSE	**Description**	**1**	**2**	**3**	**4**	**5**
1. Flow rate	Rate of flow	No flow	Trickle	Quarter depth	Half depth	Bank full
2. Visible pollutants	Pollutants at point of SUDS discharge to watercourse	No evidence of pollution	Individual small deposit/very faint odor	2-3 individual small deposits/mild odor	Many deposits, some larger, strong odor	Mass of pollution/ very strong odor
3. Pollution indicators	Evidence of watercourse response to pollution	No algae/clear waters/no gas/no odor	0-5% of surface area/occasional bubble/very faint odor	5-10% of water surface area/a few bubbles/mild odor	10-25% of water surface area/some gas/strong odor	25-50% of water surface area/ continuous bubbles/ very strong smell
4. Sediment accumulation	Buildup of silts/ sediments below outfall	No evidence of silts/ sediments	Just evident at base of structure	Covering base of structure	Up to pipe invert level	Beginning to block pipework
5. Erosion	Extent of damage caused to watercourse base/ banks by SUD outflow	No erosion	Initial evidence of erosion	Increasing levels of erosion	Erosion damage, in the future may cause bank failure	Significant erosion damage, bank failure and accretion downstream
RECEIVING WATERCOURSE ECOLOGY	**Description**					
All items	Please describe: How many of each species were viewed; density or coverage of particular plants If the species were viewed in a particular place If there was a specific feature of the SUD that seemed to be encouraging/promoting their growth					

SUDS Qualitative Observations: Explanatory Notes (Sheet 5 of 5)

COMMENTS	Description
Comments	Additional observations on the SUD system, including: Discussions with locals Discussions with site managers Progress of surrounding development Any construction activities Any addition of straw bales to suppress algae growth Degree of freezing of pond during cold spells Poor practices e.g., grass cuttings not removed from surrounding areas

4.6 Site Characterization (for Both Quantitative and Qualitative Monitoring)

Large volumes of data have been collected over the past 20 years on the performance of structural stormwater BMPs in the U.S., but the reported results may be of little use for comparing the performance efficiency of similar BMPs. A primary reason for this is the failure to report sufficient parametric data with the performance data to permit systematic analysis of the data collected (Urbonas, 1994).

To thoroughly analyze a BMP/SUDS and to understand the system and the influences on its performance, all static and state variables of the system must be known and documented. It is proposed that the collation of such information be based on the online database recently set up in the United Kingdom. (at www.suds-sites.net).

4.7 Conclusions/Approach

These protocols have been prepared to facilitate a common methodology for investigating BMP/SUDS at any location.

- Quantitative monitoring using the approach recommended in this protocol is likely to yield information on performance of BMPs/SUDS in terms of both flood alleviation and water quality improvement.
- Quantitative monitoring is likely to produce answers to only a small subset of the questions a whole life performance and costing study is expected to resolve.
- Because quantitative monitoring is extremely expensive considering the budget suggested for Phase 2, very little new information would be added to existing data sets were this project to focus only on the collection of such information.
- Quantitative monitoring should be restricted to obvious gaps in the data sets collected to date, and to collaboration with others to ensure a cost-effective approach.
- Qualitative monitoring should include observations of surrogate parameters to assess the performance of the system, and the relationship of system performance to design criteria and system maintenance. Wherever possible, qualitative observations should be supplemented by quantitative measurements.
- Consistent site characterization will be conducted and construction costs gathered for each site studied.

CHAPTER 5.0

SITE SELECTION CRITERIA AND PROPOSED MONITORING

5.1 Introduction

A number of cities, flood control districts, and environmental regulatory agencies in the United States were surveyed to gain information on the use of BMPs. The survey results (Appendix E) indicate that abundant data are available on BMP performance and costs. Since several of the organizations surveyed will be requested to provide data for detailed analysis in Phase 2 of this study, it is expected that limited, if any, quantitative monitoring of BMP performance will be conducted, and the basis for evaluating existing data will be qualitative monitoring as described in Chapter 4.0.

Although quantitative monitoring data on SUDS are less abundant than in the United States, Phase 2 monitoring in the United Kingdom is also expected to be entirely qualitative.

Chapter 4.0 describes qualitative monitoring as a cost-effective means of studying BMP/SUDS with time. The approach is not aimed at identifying the levels of specific pollutants at unique points within the system, but at evaluating the status of the system and the receiving watercourse. This evaluation can be carried on concurrently on a wide range of sites and should yield information on the following:

- The change in water quality of the BMPs/SUDS with time;
- The change in hydraulic performance of the BMPs/SUDS with time;
- The influence of maintenance on performance;
- The influence of particular design characteristics on performance;
- The impact of the system on the water quality in the receiving watercourse;
- The ecological status of the systems;
- The aesthetic benefits of the systems;
- Any risks associated with the systems; and
- Appropriate operation and maintenance schedules.

It is essential that the sites selected for Phase 2 monitoring meet two criteria:

- Useful information on the site should be readily available, and future monitoring of the performance of the site should be possible.
- There should be a proven need for research/monitoring at the site.

The needs for future research/monitoring are presented in Chapter 2.0 and in Appendix A. The monitoring discussed in this chapter is based on preliminary recommendations which identify specific goals that can be achieved within the available budget and time.

5.2 Proposed Research

The following categories have been considered in developing recommendations for future research:

- Siting Criteria;
- Design Guidelines;
- Performance (Quantitative);
- Performance (Qualitative);
- Construction Guidelines and Costs; and
- Maintenance Guidelines and Costs.

Each category is addressed in turn, with the needs and proposals indicated.

5.2.1 Siting Criteria

- Need: This is not a high priority for this project.
- Proposals: No additional research/monitoring is proposed.

5.2.2 Design Guidelines

- Need: This is not a high priority for this project
- Proposals: No additional research/monitoring is proposed
  Additional research for swales is proposed below

5.2.3 Performance (Quantitative)

5.2.3.1 Detention/Extended Detention Basins

- Need: The hydrologic and hydraulic performance of detention basins is well understood, and strategic work is not considered necessary. However, the results of recent work in Scotland indicate that the retention of pollutants in detention basins may be more substantial than previously thought.
- Proposals: A limited program is proposed to document the rate of sediment accumulation in detention basins together with the required frequency and cost of removal of such sediment.

5.2.3.2 Wet Ponds

- Need: Long-term, event-based monitoring is needed to examine the long-term changes in pollutant removal efficiencies (nutrients and heavy metals) of the wet ponds, and the potential impacts of harvesting the pond vegetation. The monitoring data would be used to determine to what extent pollutant removal rates are reduced during non-growing season and whether climatic differences have a significant impact on the detention facility performance. The nutrient buildup in vegetation must be monitored as well. Research into the rate of sediment accumulation and the increase in toxicity is also essential because these factors affect future operation and maintenance costs of SUDS.
- Proposals: In the United Kingdom, some contribution to this area is proposed through collaboration with and enhancement of existing research programs. The U.S. team will identify partners for similar work once Phase 2 commences.

5.2.3.3 Infiltration Basins

- Need: To gain greater acceptance of infiltration basins, it is necessary to determine whether temporary ponding makes clogging inevitable, and which pretreatment systems can best ensure basin longevity. Also, the performance of basins during freezing conditions should be evaluated. However, only a very small number of infiltration basins are currently in operation in the United Kingdom.
- Proposals: It may be possible to link with the monitoring proposed for a recently constructed large infiltration basin in Southern England. The U.S. team will attempt to work with the Fresno Metropolitan Flood Control District where infiltration basins have been used for many years. Since several of these basins contain water for most of the year, there will be an opportunity to assess the rate of clogging and the required maintenance and cost. Infiltration basins are also used in Long Island, New York, and the U.S. team will contact the stormwater management authorities in that area to document performance during freezing conditions.

5.2.3.4 Other Infiltration Systems

- Need: Further investigation into the potential impact on groundwater contamination for a range of soils.
- Proposals: Such investigations are likely to be expensive and are not within the budget for this project.

5.2.3.5 Filter Drains

- Need: Both the efficiency in improving water quality and the attenuation functions of filter drains need to be confirmed, as only limited data are currently available. In particular, this information is needed for operation during subfreezing conditions.
- Proposals: In the United Kingdom, some contribution to this area is proposed through collaboration with and enhancement of existing research programs. Filter drains are not used in the United States, so all Phase 2 work will be done in the United Kingdom.

5.2.3.6 Swales

- Need: The effects of different designs on pollutant removal capacity should be determined; also, whether the capacity declines with time, and whether it is influenced by the type and height of grass. Research is also needed into the effectiveness of check dams in preventing downstream migration of pollutants. Many swales operate in a fashion similar to detention basins, and the pollutant trapping aspects are also similar.
- Proposals: Such investigations are likely to be expensive, and are not within the budget for this project. In the United Kingdom, there may be an opportunity to link with a proposed monitoring program in the Midlands; however, the monitoring probably would not start until 2004.

5.2.3.7 Buffer Strips

- Need: Development of the criteria for sizing buffer strips should be a priority, as only limited empirical data are available on the minimum width or the ratio of buffer to tributary area. The optimal type and height of grass for pollutant removal, and whether its pollutant removal capacity declines with age, should be established. There is little need to study buffer strips in the United Kingdom.
- Proposals: Such investigations are likely to be expensive, and are not within the budget for this project.

5.2.3.8 Permeable Pavement

- Need: Compilation of water quality data covering seasonal changes and changes with use must be a priority. It is important to monitor frequently used car parks, such as those at a supermarket. Hydraulic monitoring should be used to confirm the attenuation within the subsurface fill.
- Proposals: Some contribution to this need area in the United Kingdom is proposed through collaboration with and enhancement of existing research programs.

5.2.4 Performance (Qualitative)

- Need: This is considered a high priority for this project. It is a cost-effective means of assessing the long-term performance of a wide range of sustainable drainage sites, as well as identifying the factors that influence performance. The relationships between maintenance, design characteristics, and construction methods and performance are also crucial.
- Proposals: It is proposed that the Phase 2 budget be allocated entirely to qualitative monitoring of a wide range of sites across the United States, Scotland, England, and Wales. This aspect of monitoring will be integrally linked with the proposals covered in the items below.

5.2.5 Construction Guidelines and Cost

- Need: This is considered a high priority for this project. In the United States, a reasonable range of cost estimates has already been compiled. However, there is no information regarding the significant variance in, for example, cost per acre drained. In the United Kingdom, cost information is available for only a small number of studies. There is no requirement through the development process for organizations to specifically cost the drainage components, and very often such costs are absorbed into the overall site infrastructure or landscaping costs.

- Proposals:
  Where sites have been identified for the estimates, it may be possible to revisit these sites and to identify the factors contributing to the cost levels.
  - Organizations may be willing to release some details, although for systems that have been in place for a few years, the costs may be too difficult to retrieve. The costs for new facilities that are under construction may be more readily accessible. A number of projects have been identified, mostly in Scotland, whose developers have indicated a willingness to participate in collating both construction and maintenance cost information.
  - Though detailed qualitative surveys of a wide range of sites, specific characteristics of each scheme can be appraised, and the likely contribution of different characteristics to total cost can be estimated. Such characteristics are likely to include inlet/outlet infrastructure, construction methods/materials, vegetation and planting, pipework, erosion control, and pretreatment systems (including sediment management), whose variability may be driven by regulatory requirements, site geology/topography, site use, and development value.
  - Obvious links between performance and specific design characteristics will highlight deviations from best practice, or weakness or lack of current guidance.

5.2.6 Maintenance Guidelines and Cost

- Need: This is considered a high priority for this project. Most of the common-sense recommendations for operation and maintenance regimes are derived from a common source, and are likely to have little empirical basis. There is an urgent need to appraise the frequency and cost (level of activity) of maintenance required to achieve appropriate performance levels of BMP/SUDS in different climates.
- Proposals: A range of sites will be visited on a regular basis to establish the effectiveness of the operation and maintenance strategies in terms of achieving consistent, long-term performance. Any correlation between maintenance and performance will be identified, and the causes of performance failures will be investigated.

5.3 Site Selection Criteria and Consultation

5.3.1 Quantitative Monitoring

In the unlikely event quantitative monitoring is required, the site selection criteria were defined as follows:

- Research is ongoing at the site, and additional work is therefore possible at relatively low cost. This will ensure that maximum value is achieved for the funds available.
- Developers, responsible organizations, and stakeholders are prepared to fully or partially fund the monitoring.
- The site is in reasonable proximity to the research contractor base.
- Research need has been established.

5.3.2 Qualitative Monitoring

The site selection criteria for qualitative monitoring were defined as follows:

- Developers, responsible organizations, and stakeholders are interested in promoting and developing BMPs/SUDS and in participating in the project.
- The monitoring covers a range of components or systems.
- The design characteristics for each component can be established.
- The site is in reasonable proximity to the research contractor base.
- The site is readily accessible for inspection.
- Research need has been established.

5.4 Consultation and Site Selection in the United Kingdom

In Scotland, a database of more than 700 SUDS sites has been compiled by SEPA. Researchers at the University of Abertay, Scotland, are fully aware of the range of sites that would be suitable for further study, as a result of ongoing research projects for SEPA, SNIFFER, and Scottish Water. Therefore, no additional consultation was required.

In England and Wales, a large number of organizations were consulted in an attempt to identify SUDS sites that would meet the above criteria, including the following organizations:

- Local Authorities
  - Arun District Council
  - South Gloucestershire Council
  - Borough of Telford & Wrekin
  - Leeds City Council
  - Wakefield District Council

- Water Companies
  - Severn Trent Water
  - Yorkshire Water
  - Thames Water
  - Southern Water

- Environment Agency
  - All regional and area SUDS representatives

- Universities
  - Imperial College
  - Coventry University
  - Bradford University
  - Middlesex University

- Developers
    - Peabody Trust
    - Westbury Homes
    - Bryant Homes
    - English Partnerships
    - Elvetham Heath

- Designers
    - Ove Arup Consulting Engineers
    - Peter Brett Associates
    - White Young Green

- Others
    - CIRIA
    - Robert Bray Associates
    - Formpave

This consultation was undertaken in conjunction with the development of an on-line SUDS database, funded by the Environment Agency of England & Wales. This database, which can be viewed at /www.suds-sites.net/, incorporates the following categories of information:

- Level 1
    - Site location
    - Site status
    - Site description
    - Site SUDS components
    - Drainage objectives
    - Maintenance responsibilities
    - Site acceptability
    - Site performance
    - Site monitoring
    - Relevant contacts

- Level 2
    - Design details
    - Infiltration details
    - Types and frequency of maintenance
    - Capital costs
    - Maintenance/rehabilitation costs
    - Social issues
    - Health and safety issues

5.5 Proposed Monitoring Sites in the United Kingdom

5.5.1 Quantitative Monitoring

In order to maximize the value from any budget allocated to quantitative monitoring, the key criterion was that the sites must be already part of a research campaign.

A number of partner research projects have been identified in the United Kingdom, and it is proposed that the limited funds available for this type of monitoring be directed to increasing the output from the partner projects. Relevant partner projects are listed below:

- Bradford University Ph.D. student research on wet ponds.
- Coventry University M.Sc. student research on swale monitoring.
- DTI—ongoing HR Wallingford monitoring study (retention pond, wet swale, permeable pavement).
- Proposed monitoring at Angmering Infiltration Basin, funded by developers
- Environment Agency monitoring campaigns (e.g., ongoing monitoring at Wokingham).
- Ph.D. program at Herriott Watt University funded by Scottish Water (whole-life performance of SUDS).
- Ph.D. program at University of Abertay (Performance of filter drains).
- Ongoing program at Edinburgh University funded by Scottish Water (Sediments in ponds and detention basins).
- Holistic assessment of treatment trains at DEX, Dunfermline, funded by Wilcon Homes (Scotland) Ltd.

The sites proposed for study in this project are listed in Table 5-1.

Table 5-1. Proposed Quantitative Monitoring Sites (U.K.).

SUD Component	Proposed Site	Monitoring Proposed	Comments
Wet Ponds	Welcome Break Service Station, Hopwood	Additional event-based monitoring. Sediment monitoring. Possible vegetation monitoring.	Link with ongoing HR Wallingford monitoring at this site.
	Thorpe Business Park, Leeds	Additional event-based monitoring. Sediment monitoring. Possible vegetation monitoring.	Link with ongoing HR Wallingford/Bradford University monitoring at this site.
	Wakefield Ponds	Additional event-based monitoring. Sediment monitoring. Possible vegetation monitoring.	Link with Ph.D. student research at Bradford University.
	DEX (Dunfermline) Stenton, Glenrothes Newbridge, Edinburgh	Wide-ranging water, biology, costs, and social.	Continues existing studies at DEX with little extra cost.
Infiltration	Roundstone Lane, Angmering	Transfer of pollutants into subsoil. Infiltration capacity of subsoil, and impact of ponding. Effectiveness of pretreatment system.	Scope dependent on monitoring finally proposed by developers.
Detention Basins	DEX, Dunfermline St. Johnstone FC, Perth Ardler, Dundee Emmock Woods, Dundee	Hydrology of two ponds and sediment accumulation.	Links in with longer study at DEX.
Other Infiltration Systems	None proposed.		
Filter Drains	Welcome Break Service Station, Hopwood	Some event-based water quality monitoring.	Link with ongoing HR Wallingford monitoring at this site.
	Broxden, Perth Walker Dam, Aberdeen Further sites available	Hydrology for one year and targeted sonde monitoring.	Addresses known.
Swales	Sherwood Energy Village DEX Spine Road, Dun'ne Emmock Woods, Dundee West Grange, Dundee Ardler, Dundee Gallows Knowe, Dunning Clarverhouse Ind., Dundee McNabb Avenue, Montrose Panmuirfield, Dundee	Unable to confirm at present.	Link with Coventry University research (NB research timescale may not work).

SUD Component	Proposed Site	Monitoring Proposed	Comments
Buffer Strips	Welcome Break Service Station, Hopwood	Some event-based monitoring may be possible.	Link with ongoing HR Wallingford monitoring at this site.
Permeable Pavements	Tesco Superstore, Kidderminster Royal Bank of Scotland, Edinburgh Tesco Duloch Park NATS Edinburgh Tesco Forfar		

The final scope of the monitoring program will be determined after a full budget review for Phase 2 of the study.

5.5.2 Qualitative Monitoring

It is proposed that between 20 and 30 sites in England and Wales and approximately 30 sites in Scotland be selected for qualitative monitoring in Phase 2. The same sites will be used to obtain construction details, construction cost information, maintenance schedules and reviews, as detailed earlier in this chapter. Sites available for possible selection are presented in Appendix F, and summarized in the table below. The spread of sites reflects the frequency of use of different components.

Table 5-2. Potential Qualitative Monitoring Sites.

	Sites in England	Sites in Scotland
Swales	10	12
Permeable pavements	11	4+
Filter drains	4	10
Filter strips	1	0
Soakaways	7	3
Retention ponds/wetlands	15	7
Detention basins	4	11
Total	52	47

Supportive research/funding has also been identified for this element of the project, as detailed below:

- HR have funding for a Marie Curie Research Fellow starting in April 2003, funded by the EC.
- Work at DEX, Dunfermline funded by Wilcon Homes (Scotland) Ltd.
- MSc project at Abertay on Whole Life Costs.
- Determination of costs of eight SUDS components funded by Scottish Water.

5.6 Selection of U.S. Organizations for Involvement in the Phase 2 Study

Identification of U.S. organizations that may participate in the Phase 2 study involved the following steps:

- A survey of U.S. cities, flood control districts, environmental regulatory agencies, and other organizations that might have data on BMP performance and costs or other useful information;
- Development of selection criteria; and
- Review of the survey data and application of the selection criteria.

5.6.1 Survey of Organizations

A survey questionnaire was developed and sent to more than 80 organizations. The list of organizations and contact persons (Appendix E) was compiled based on knowledge of the project team, specialty conference mailing lists, and other sources. Of the approximately 50 organizations that responded, 31 appeared to have data applicable to this study (Appendix D). All but one of the 31 organizations were using two or more of the 12 BMPs discussed in this study. The exception is the Occoquan Watershed Monitoring Laboratory, which is not responsible for operating any BMPs, but has conducted an extensive monitoring program for a number of years. Twelve organizations indicated that they had capital and operation and maintenance cost data; 15 stated that they were conducting routine monitoring of their BMPs, and 14 organizations indicated that they were conducting assessments of conditions downstream from their BMPs.

5.6.2 Selection Criteria

The following criteria were used to identify the group of organizations that potentially could participate in the Phase 2 study. It was not anticipated that each organization would meet all the criteria, but that collectively they would provide information important to this study.

- The organizations would represent a wide range of geographic locations in the US.
- The groups would comprise organizations that have studied the 12 study BMPs. Organizations that have constructed large numbers of BMPs would be favored.
- The organizations would be in a position to provide information on BMP capital and operation and maintenance costs.
- The organizations would have conducted extensive monitoring programs, including downstream biological assessments.
- The group would be capable of meeting the research needs identified in Chapter 2.0.

5.6.3 Potential Phase 2 Organizations

The following organizations constitute a group that was considered to best meet the selection criteria. They are not listed in any particular order of importance.

- City of Austin, TX
- Prince Georges County, MD
- Fresno Flood Control District, CA
- Denver Urban Drainage and Flood Control District, CO
- Montgomery County, MD
- California Department of Transportation (CALTRANS)
- South Florida Water Management District
- Occoquan Watershed Monitoring Laboratory, University of Virginia
- King County, WA

It should be noted that this is a preliminary list based on the information available from the survey. To date, none of the organizations listed has agreed to participate in Phase 2. A more detailed evaluation of their data and the final selection will be made at the beginning of the Phase 2 study.

Chapter 6.0

FINDINGS, CONCLUSIONS, AND RECOMMENDATIONS

This project is being conducted in two phases. Phase 1—the subject of this report—encompasses the following elements:

- A literature review and a survey of stormwater authorities and organizations in the United States and the United Kingdom to identify the most commonly used BMPs/SUDS;
- Establishment of protocols for whole-life costs and performance data for BMPs/SUDS; and
- The selections of a limited number of BMPs/SUDS stormwater authorities and organization for quantitative or qualitative monitoring and further assessment during Phase 2.

6.1 Findings and Conclusions of the Literature Review

The literature review identified 243 publications that report results of studies for 11 types of BMPs/SUDS. The literature review revealed that despite the amount of research and study conducted over the last 25 years, significant gaps remain in our knowledge about the performance and whole life costs of BMPs/SUDS. There is great variability of performance data and only a limited common framework. Phase 2 of this study will be designed to fill these gaps by working with agencies and owners responsible for implementation and maintenance of these facilities to develop better performance and cost data.

The key issues and research needs highlighted by the literature review may be summarised as follows:

6.1.1 Dry Extended Detention Basins

Dry extended detention basins are vegetated depressions, which are normally dry except during and immediately following storm events when they provide temporary storage for stormwater runoff, allowing the solids particles and attached pollutants carried in the runoff to settle out. Those facilities are widely used in both the United States and the United Kingdom.

They have few siting requirements and little headloss, which make them adaptable to a variety of settings. Most published reports document moderate removal of particulates and associated pollutants in such basins, but only very limited removal of nutrients and other dissolved constituents. Although maintenance guidelines emphasize the need for periodic sediment removal and clearing of the basin's outlet structure, the majority of maintenance consists of vegetation management, mostly grass mowing. The frequency of mowing depends on aesthetic considerations rather than performance needs.

Siting criteria for extended detention basins are well established. Pollutant removal appears to be dependent primarily on draining time and, to a lesser extent, on basin geometry and depth. Extended detention basins have been the subject of numerous monitoring studies; consequently, little additional meaningful data is likely to be gained from devoting the limited resources of this project to quantitative monitoring. There is a larger uncertainty in construction costs, with wide ranges reported by various sources. An area where additional information is needed is the minimum level of maintenance needed for proper functioning.

6.1.2 Retention Ponds and Wet Ponds

Retention ponds and wet ponds are widely used in both the United States and the United Kingdom. Ponds are one of the best means of incorporating the multipurpose SUDS approach of water quality, amenity, and habitat. Results of previous monitoring have indicated high efficiency for solids removal and moderate to low efficiency for removal of dissolved constituents. Much of the difference in removal efficiency among the sites reported in various studies may be attributable to differences in the water quality of the permanent pool. The wet weather discharge from ponds, especially during smaller storm events, consists primarily of displaced permanent pool water, which strongly affects the calculated removal efficiencies.

Research needs for ponds include the following:

- Siting criteria for safety and acceptance;
- Data on permanent pool size relative to the design water quality storm;
- Information on the role of emergent vegetation in pollutant removal;
- The impact of vegetation harvesting on long-term nutrient reduction; and
- Long-term maintenance requirements.

6.1.3 Infiltration Basins

Infiltration basins are not widely used in the United Kingdom but are very common in some parts of the United States, such as the California central valley and Long Island, New York. They are especially attractive because of their potential for helping maintain the pre-development hydrology by reducing the volume of stormwater runoff. Unfortunately, infiltration basins have a history of high failure rates when sited in areas with marginal soil and groundwater conditions.

Research needs for infiltration basins include the following:

- Siting criteria because of perceived concerns about groundwater quality impacts and early hydraulic failure;
- Ways to minimize clogging;

- Performance in freezing weather; and
- The quality of water below basin.

6.1.4 Infiltration Trenches and Soakaways

Infiltration trenches and soakaways are relatively common in the United Kingdom, but less so in the United States. Like infiltration basins, they are desirable for maintaining the predevelopment hydrology. Unlike the basins, they are installed below grade and are therefore unobtrusive. They also have a history of high failure rates.

Research needs for infiltration trenches and soakaways include the following:

- Siting criteria;
- Design guidance;
- Data on groundwater quality;
- Construction cost; and
- Expected life.

6.1.5 Retention/Irrigation Systems

Retention/irrigation systems are designed to promote infiltration of runoff by irrigation in areas where the soils are not permeable enough for infiltration basins and trenches. There is no published information on the performance of such systems, and their only known application is environmentally sensitive watersheds in the Austin, Texas, area where stormwater discharges are strongly discouraged.

Because of the lack of published studies and their limited use, additional research into retention/irrigation systems is needed in the following areas:

- Sizing of irrigation area;
- Optimum rate of runoff application;
- Effect on vegetation health of irrigation during wet weather;
- Construction costs;
- Maintenance frequency for pumps and irrigation system; and
- Cost of operation.

6.1.6 Vegetated Swales

Vegetated swales are grass-lined channels designed for stormwater conveyance and treatment. The swales are common in both the United States and the United Kingdom, although they are often not engineered to achieve optimal pollutant removal. Although the siting criteria for swales are well established, they can be difficult to implement in urban areas where they may conflict with land development or zoning codes that require the use of curbs and gutters. Monitoring them for water quality is often difficult because of the absence of a single inflow point.

Research needs for vegetated swales include the following:

- Minimum duration of hydraulic residence time (current recommendations are all derived from a single study);
- Effect of grass type and height on removal of pollutants;
- Construction costs; and
- Maintenance recommendations.

6.1.7 Vegetated Buffer Strips

Vegetated buffer strips are grassy areas where runoff is treated as diffuse overland flow. There is abundant information on these facilities in agricultural publications; however, only six U.S. publications on their use for urban runoff were identified. Vegetated buffer strips are relatively common in the United States; however, they are usually not designed for optimal treatment of runoff. One example is highway shoulders. Vegetated buffer strips receive relatively little recognition for their performance but are generally more robust and less likely to become channelized than swales.

Research needs for buffer strips include the following:

- Sizing criteria relative to drainage area;
- Effect of grass type and height on pollutant removal efficiency; and
- Cost data.

6.1.8 Sand Filters and Organic Filters

Sand filters and organic filters are widely used in Austin, Texas, and in several other areas of the United States, but they are not used in the United Kingdom. There are a wide variety of configurations and media types, and high removal rates of solids and associated pollutants have been reported. Because runoff passes through the filters at a relatively low rate, these facilities provide substantial attenuation of peak runoff. The siting, design, and performance of sand filters are well documented; however, there is little information on maintenance frequency and its relationship to solids loadings. Maintenance costs are another area where additional data are needed.

6.1.9 Water Quality Inlets and Oil/Water Separators

Water quality inlets and oil/water separators were the subject of the 11 U.S. publications identified in the literature review. Water quality inlets, such as catch basins, are very common in the United States, especially in the Northeast, and gullypots are common in the United Kingdom. Most of the available data indicated that they are not very effective for water quality improvement, and no further research into these devices is recommended for this project.

6.1.10 Bioretention Areas

Bioretention areas were the subject of the 11 U.S. publications reviewed. These facilities combine many of the aspects of infiltration (trenches) and filtration (sand filters) technologies. A relatively new development with few known installations, bioretention areas are expected to achieve high rates of pollutant removal. By substantially reducing the volume and rate of stormwater runoff, they help maintain the predevelopment hydrology. Priority areas for research are long-term performance and routine maintenance requirements.

6.1.11 Porous and Permeable Pavements

Porous and permeable pavements have been installed as demonstration projects; although their use is not widespread, they are beginning to gain popularity in the United Kingdom. Available data indicate that porous and permeable pavements achieve significant attenuation of peak flows and generally high pollutant removal.

Research needs for porous and permeable pavements include the following:

- Structural integrity issues;
- Performance of saturated subgrades;
- Performance under load;
- Effects on water quality;
- Design guidelines;
- Operating life; and
- Maintenance requirements to preserve permeability.

6.1.12 The Impact of BMP/SUDS on Downstream Ecology

The impact of BMP/SUDS on downstream ecology was the subject of 13 publications from the United States and two from the United Kingdom, which were reviewed as part of this project. Although there is extensive documentation of the adverse downstream impacts of urbanization, little has been published about the positive effects of BMPs/SUDS. Studies dealing with this subject are extremely resource-intensive, and it would not be feasible to develop additional information in this area with the limited funds available for this study.

6.2 Findings and Conclusions for Performance Protocols

Numerous studies have been conducted, particularly in the United States, with the objective of characterizing the pollutant reduction associated with each type and design of stormwater facility. Consequently, monitoring of this type is not a primary objective of this study. More urgently needed are qualitative monitoring protocols for evaluating the performance of BMP/SUDS by using empirical observations to determine whether a structural or a maintenance-related failure has occurred. These protocols will allow stormwater managers to determine, relatively inexpensively, the current status of their stormwater facilities. Suitable monitoring protocols have not been available to date.

6.3 Findings and Conclusions for Whole-Life Costing Protocol

Reported construction costs of BMP/SUDS vary widely, and little is known about the factors that contribute to higher costs. The information about the maintenance of BMP/SUDS is equally limited, even though the long-term maintenance costs can be an important component of the whole life cost of the facility. To correct this deficiency, the Whole-Life Costing Protocol has been developed and will be provided as a separate document.

6.4 Recommendations for Phase 2

Because of a limited budget, only qualitative monitoring will be conducted during Phase 2 of the project. The qualitative monitoring will be carried out using the protocols developed in Phase 1. When possible, cost data will be collected to complete development of the Whole-Life Costing Protocol.

In the United States, significant quantitative monitoring data are already available, so additional quantitative monitoring is not planned. The principal tasks for Phase 2 will be to resolve the discrepancies in published data in order to establish the design parameters that have the greatest impact on performance, and to carry out qualitative performance monitoring concurrently with the monitoring done in the United Kingdom. In the United States, the project team will carry out qualitative monitoring through responsible stormwater agencies.

In the United Kingdom, the project team will conduct the qualitative monitoring. Individual sites can be identified, concentrating on areas where the data gaps can be filled at the least cost. The team will also attempt to obtain funding for the monitoring from other sources, particularly from academia.

The U.S. and U.K. teams will try to avoid duplicating the work, although there will be a need for some overlaps for benchmarking purposes. Regions of comparable topography and climate will be identified.

An important product of Phase 1 of the project is the performance monitoring protocols, which had been lacking to date. These protocols should be made available to monitoring organizations to help build larger comparable datasets. Websites are envisioned as the most appropriate medium to post the protocols and to receive and review feedback.

6.5 Phase 2 Work Plan

It is recommended that Phase 2 of the project include the following elements:

- Refinement and testing of the Whole-Life Costing Protocol;
- Compilation of existing data held by stormwater agencies, municipalities, and other organizations;
- Qualitative monitoring;
- Assessment of pollutant removal performance;
- Assessment of hydraulic performance; and
- Assessment of BMP/SUDS maintenance practices.

APPENDIX A

LITERATURE REVIEW OF BMPs/SUDS

TABLE OF CONTENTS

LIST OF TABLES

LIST OF FIGURES

CHAPTER 1.0

INTRODUCTION

This report presents a summary of the literature concerning siting, design, cost, performance, and operational aspects of best management practices (BMPs) and sustainable urban drainage systems (SUDS). It is important to distinguish between the terms BMP and SUDS and to establish the differing management goals for their implementation.

The concept of a sustainable drainage system was developed by the Sustainable Urban Drainage Scotland Working Party (SUDSWP, 1999), based upon the idea of an integrated stormwater management plan that addresses pollutant reduction and flood control, while providing wildlife habitat and aesthetic amenities. This holistic, multipurpose approach was illustrated by D'Arcy and Frost (2001) and is reproduced as Figure 1-1.

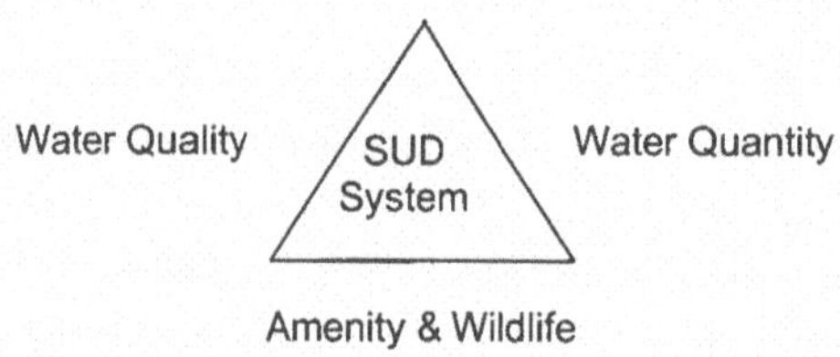

Figure 1-1. Sustainable Drainage Triangle Concept. Reprinted from *Science of the Total Environment,* Vol. 265, D'Arcy, B.J., and Frost, C.A., The Role of Best Management Practices in Alleviating Water Quality Problems Associated with Diffuse Pollution, pp 359-367 (2001), with permission from Elsevier.

As D'Arcy and Frost (2001) note, the concept recognizes that the needs of flood control engineers and pollution control scientists must be considered together if the environment is to be adequately protected. If the resulting facilities are to be acceptable for adoption and maintenance, they must be attractively landscaped and integrated into the local community as amenity green space. Consequently, a SUDS can be defined as consisting of structural elements to address pollution and flooding from urban stormwater runoff in an attractive, natural way. The concept of sustainable drainage systems has subsequently become an important consideration across Europe (Chocat, Barraud, and Alfakih, 2001).

The concept of sustainable drainage systems is quite different from the concept of best management practices which is a central component of the U.S. Clean Water Act (CWA, 1972 and subsequent amendments). The CWA requires the use of measures (practices) to reduce the pollutants in discharges from industrial facilities as well as in urban runoff. These practices may be either structural or nonstructural. An example of a nonstructural practice would be a campaign to educate the public to the potential impact of dumping hazardous materials into storm drains. Structural BMPs, on the other hand, consist of ponds, swales, or other constructed features designed specifically for pollutant removal. This focus on quality over quantity mainly occurs because the U.S. Environmental Protection Agency (EPA) does not have any authority in the area of flood control despite the obvious environmental impacts of the larger runoff volumes and rates caused by urbanization. The design of BMPs installed in the United States is driven by performance standards to achieve certain levels of pollutant removal; consequently, little attention is given to aesthetic values, and BMPs are sited behind buildings, fences, or other features to hide them from view.

Another significant difference between the U.S. and U.K. approaches is the configuration of existing drainage systems in urban areas. For many years drainage systems that combined domestic and industrial wastewater with stormwater runoff were commonly used in Europe. These combined flows were discharged without treatment into receiving waters. Council Directive 91/271/EEC (May 1991) changed the ground rules by requiring secondary treatment for urban wastewater. Because about 70% of drainage systems in the United Kingdom are combined (Marsalek and Chocat, 2001), stormwater runoff from smaller events also receives secondary treatment. Consequently, implementation of BMPs/SUDS for water quality treatment is not as high a priority in the catchments served by the combined systems. Instead, there is an emphasis, particularly in England, on reduction of storm runoff volumes and rates to prevent combined sewer overflows (CSOs) and any associated flooding.

Nevertheless, there is a common ground between these approaches to stormwater management, which this literature review and subsequent study will address. The focus will be on stormwater management measures that are common to both the United States and the United Kingdom. Consequently, the following criteria were used to identify practices that are included in this review:

- The practice must be a structural control, such as a wet pond, swale, and so forth.
- The practice may address either water quality or quantity. Where quantity/rate control is the primary function of the practice, the emphasis will be on structures directed at smaller, more frequent events (i.e. less volume than the 2-year, 24-hour event).
- Although on-site practices may be included in the review, the focus will be on practices appropriate for 0.5 ha (1.2 ac) and larger catchments.

Using these criteria, the following structural controls were identified for review:

- Extended Detention Basins
- Wet ponds/Constructed Wetlands/Retention Basins
- Infiltration Trenches/Soakaways/Filter Drains
- Infiltration Basins
- Retention/Irrigation
- Sand and Organic Filters
- Vegetated Swales
- Vegetated Buffer Strips
- Porous Pavement
- Bioretention
- Water Quality Inlets (enhanced catch basins).

Each technology is discussed in a separate chapter, and the review is based initially on the data contained in EPA's menu of BMPs (http://cfpub.epa.gov/npdes/stormwater/menuofbmps/menu.cfm).

An important consideration in the implementation of BMPs/SUDS is the degree to which ecological integrity of stream segments is protected. This is a function not only of improvement in water quality associated with a particular BMP, but also the degree to which the predevelopment hydrology is preserved. Consequently, a brief review of the literature associated with ecological impacts has also been conducted.

CHAPTER 2.0

DETENTION BASINS/ EXTENDED DETENTION BASINS

2.1 Description

Dry detention basins are generally vegetated depressions which are dry except during and immediately following storm events. (Wet detention basins, which have a permanent pool of water throughout the year, are discussed in Chapter 3.0.) During a storm, surface runoff is routed through the detention basin and the outlet is restricted so that the basin fills with runoff. Dry detention basins are normally used to reduce maximum runoff rates from developed areas to their pre-development levels. The control of the maximum runoff rates serves to protect drainage channels below the basin from erosion and to reduce downstream flooding. Figure 2-1 shows a detention basin used to treat highway runoff at the DEX site in Scotland.

Figure 2-1. Dry Detention Basin, DEX Site, Scotland.

Dry extended detention ponds are basins whose outlets have been designed to detain the stormwater runoff from a water quality design storm for some minimum time, such as 48 hours. Extending the detention times beyond what is required for flood control improves water quality by allowing coarse silts and associated pollutants to settle (CIRIA, 2000a,b). Such systems may contain a low flow channel and a small permanent pool as a landscape feature, and to improve ecological benefits. A schematic of an extended detention basin with a micropool is shown in Figure 2-2.

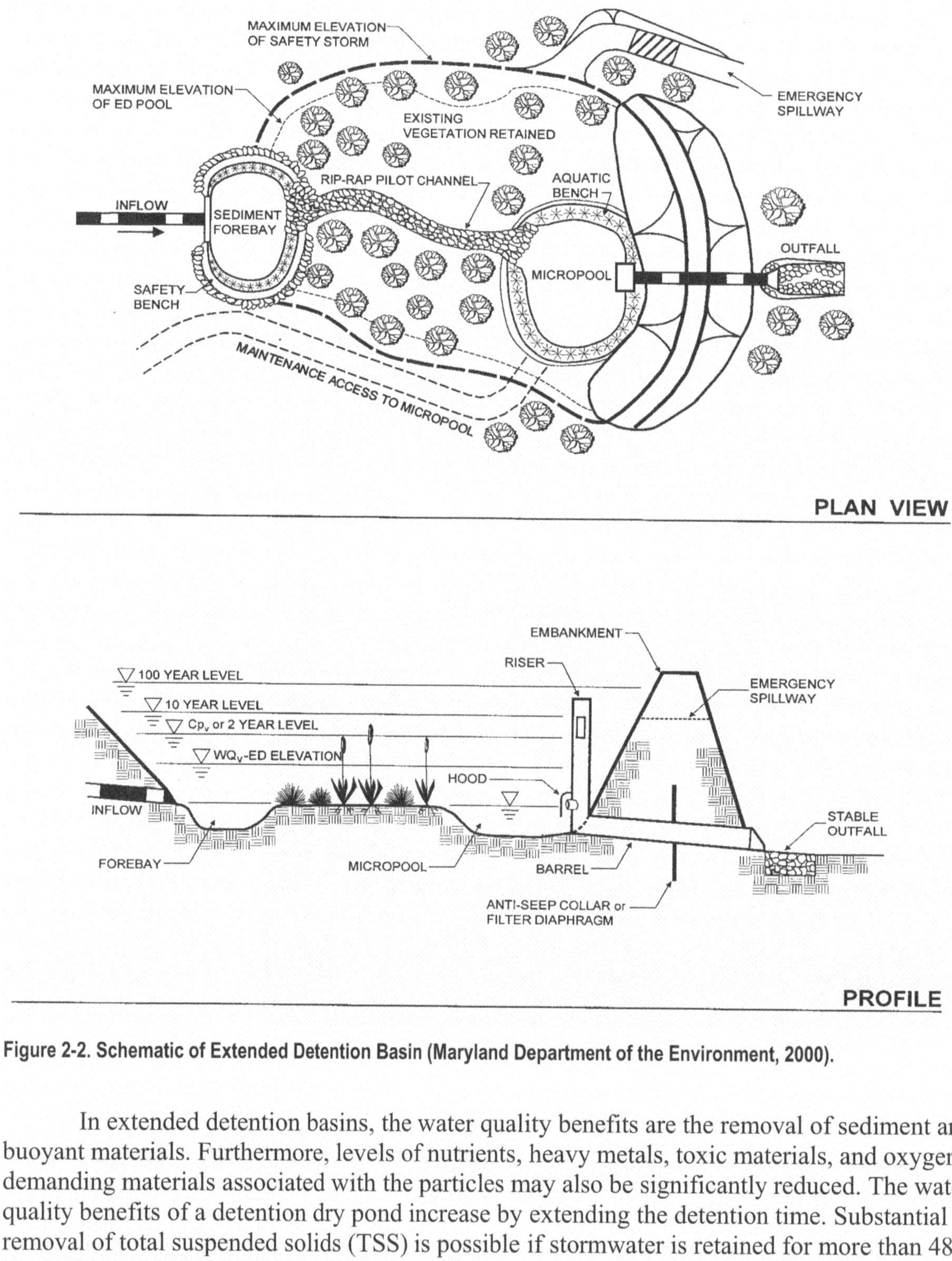

Figure 2-2. Schematic of Extended Detention Basin (Maryland Department of the Environment, 2000).

In extended detention basins, the water quality benefits are the removal of sediment and buoyant materials. Furthermore, levels of nutrients, heavy metals, toxic materials, and oxygen-demanding materials associated with the particles may also be significantly reduced. The water quality benefits of a detention dry pond increase by extending the detention time. Substantial removal of total suspended solids (TSS) is possible if stormwater is retained for more than 48 hours. Detention facilities frequently are employed for temporary sediment control during construction, and it may be possible to retain some of these installations permanently (Young et al., 1996).

Dry basins can be considered where the retention of large volumes of water is either not beneficial or inappropriate. It is likely that they are easier to maintain than wet ponds and may be considered less of a health and safety hazard, although the risks must be assessed for wet, dry, and filling conditions. Dry detention areas can be integrated with, for example, sports fields, thus allowing a dual use where land availability is limited. Consideration should be given to their use in combination with other controls such as grassy swales and vegetated filter strips to achieve required TSS removal.

Although detention facilities designed for flood control have different design requirements than those used for water quality enhancement, it is possible to achieve these two objectives in a single facility. For example, the City of Austin has a dual-purpose facility on Great Northern Blvd.

Although most detention basins are constructed as surface depressions, land costs are driving the use of basins located below grade, underneath parking lots or other structures. These may be constructed on corrugated pipe or of proprietary plastic cells available from a number of manufacturers (e.g., Wavin Plastics Limited in the U.K. or Invisible Structures, Inc., in the U.S.).

2.2 Advantages

- Because of their simple design, extended detention basins are relatively easy and inexpensive to construct and operate;
- Extended detention basins can capture substantial amounts of sediment and the toxics fraction associated with particulates; and
- Widespread use of detention basins with sufficient capture volume can provide significant control of channel erosion and enlargement in a watershed.

2.3 Limitations

- Limitation of the diameter of the orifice may not allow use of extended detention on watersheds of less than 2.0 ha (5 ac), because it may require very small orifice that would be prone to clogging.
- Differential elevation is necessary between inlet and outlet.
- Improper design or construction may result in a mud hole that becomes a nuisance to local residents and an environmental hazard.
- Detention basins are not practical for drainage areas over 100 ac.
- Freeboard is required between spillway and crest of embankment.
- Dry extended detention ponds provide only moderate pollutant removal when compared to wet extended detention ponds and other structural stormwater practices, and they are ineffective at removing soluble pollutants. Although wet ponds can increase property values, dry ponds can actually detract from the value of a home.
- Dry extended detention ponds on their own provide only peak flow reduction and do little to control overall runoff volume, which could result in adverse impacts downstream.
- Rutting and damage caused to grass and upper soil structure by heavy vehicles (CIRIA, 1993) may result in reduced bearing capacity of the soil.

- If detention storage volume is exceeded during a storm event, pollutant removal may be decreased providing only partial treatment of runoff (CIRIA, 1993).

2.4 Siting Criteria

Dry extended detention ponds are among the most widely used stormwater management practices in the United States and are especially appropriate for retrofit situations where their low hydraulic head requirements allow them to be sited within the constraints of the existing storm drain system. Although dry extended detention ponds can be applied rather broadly, designers need to ensure that they are feasible at the site in question. This section provides basic guidelines for siting dry extended detention ponds.

In general, dry extended detention ponds should be used on sites with a minimum area of 2.0 ha (5 ac). On smaller sites, it can be challenging to provide channel or water quality control because the orifice diameter at the outlet needed to control relatively small storms becomes very small and thus prone to clogging. In addition, it is generally more cost-effective to control larger drainage areas due to the economies of scale.

Extended detention basins can be used with almost all soils and geology, with minor design adjustments for regions of karst topography or in rapidly percolating soils such as sand. In these areas, extended detention ponds should be designed with an impermeable liner to prevent ground water contamination or sinkhole formation.

Except for the case of hot spot runoff, the only consideration regarding ground water is that the base of the extended detention facility should not intersect the ground water table. A permanently wet bottom may become a mosquito breeding ground and a health and safety risk. Research in Southwest Florida (Santana et al., 1994) demonstrated that intermittently flooded systems, such as dry extended detention ponds, produce more mosquitoes than other pond systems, particularly when the facilities remained wet for more than three days following heavy rainfall.

A stormwater retrofit is a stormwater management practice (usually structural) put into place after development has occurred to improve water quality, protect downstream channels from erosion due to increased runoff velocities, reduce flooding, or meet other specific objectives. Dry extended detention ponds are very useful stormwater retrofits, and they have two primary applications as a retrofit design. In many communities in the past, detention basins have been designed for flood control. It is possible to modify these facilities to incorporate features that encourage water quality control and/or channel protection. It is also possible to construct new dry ponds in open areas of a watershed to capture existing drainage.

A study in Prince Georges County, Maryland, found that stormwater management practices can increase stream temperatures (Galli, 1990a). Overall, dry extended detention ponds increased temperature by about 3°C (5°F). In cold water streams, dry ponds should be designed to detain stormwater for a relatively short time (i.e., less than 24 hours) to minimize the amount of warming that occurs in the practice.

2.5 Design Guidelines

Estimating the appropriate dimensions of a BMP facility is largely based on a trial and error process in which the designer tries to fit the required BMP volume so that it works well with the site. Each site has its own unique limiting factors. Constraints other than the existing topography include, but are not limited to, the location of existing and proposed utilities, depth to bedrock, and location and number of existing trees. The designer can analyze possible basin configurations by varying the surface area and depth and then determining the corresponding available storage (Young et al., 1996).

Most design guidelines in the United Kingdom are based on CIRIA (1993), which confirms that most cases will be site specific, requiring any design problems to be considered on an individual basis. The ultimate design question for flood control is the storage volume required to achieve the limit of outflow and the duration of the worst-case critical storm scenario (Butler and Davies, 2000).

Many of the design criteria for extended detention basins are not well supported by empirical monitoring. The criteria believed to be the most critical include time to drain from basin full conditions, basin geometry, and water depth. Although it is common sense that a longer detention time will allow settling of smaller particles, no well-designed study has investigated this relationship under field conditions. Studies have been conducted of basins with different drain times, but these facilities also differ in geometry, influent water quality, and design depth. An optimum study would monitor a single facility, changing the outlet orifice periodically to achieve different drain times. This would allow a relationship to be developed between drain time and pollutant removal and indicate how sensitive performance is to this parameter.

Likewise, a minimum length to width ratio is often specified in design guidelines to improve performance and reduce short circuiting within the basin. The requirement for a longer flow path is often based on the model of a wastewater clarifier, which operates continuously instead of detention basins that operated more as a batch process. Finally, minimum and maximum depths are often specified in guidelines to improve sedimentation (maximum depth) and reduce resuspension (minimum depth). As with the other factors, no literature that evaluated these parameters was identified.

Despite these uncertainties, U.S. and U.K. guidelines are generally similar but for the more rigid requirements in the United States. Typical criteria are listed in Table 2-1.

Table 2-1. Design Guidelines.

Design Criteria	Current Guidance/Recommendations
Hydraulic Residence Time	♦ Extended detention facilities should be sized to completely capture the water quality volume and this volume should be increased by a factor of 10% to accommodate reductions in the available storage volume due to deposition of solids in the time between full-scale maintenance activities.
Planting	♦ The landscape design should specify appropriate grass species (to give a dense sward) and plants that will grow well on the site and in the local soils. The choice of species must take into account the fact that the vegetation will be subject to periodic inundation and water flow (CIRIA, 2000 a,b).
Side Slopes	♦ The maximum side slope should be designed to meet safety and maintenance requirements.
Basin Depths	♦ Basin depths optimally range from 0.6-1.5 m (2-5'). In the U.K., maximum basin depths should not exceed 3 m (10') to avoid damaging the vegetation (CIRIA, 2000a,b).
Low Flows	♦ A channel should be provided for dry weather flows.
Outlet Infrastructure	♦ The facility's drawdown time should be regulated by a gate valve or orifice plate located downstream of the primary outflow opening. ♦ In general, the outflow structure should have a trash rack or other acceptable means of preventing clogging at the entrance to the outflow pipes. The outflow structure should be sized to allow complete drawdown of the water quality volume in 48-72 hours with no more than 50% of the water quality volume draining from the facility within the first 24 hours (U.S. recommendations). ♦ U.K. guidance recommends a total emptying time on the order of 24 hours. The outflow structure should be fitted with a valve (or penstock) so that discharge from the basin can be halted in case of an accidental spill in the watershed. This same valve can also be used to regulate the rate of discharge from the basin. ♦ By using a multistage outlet, flows in excess of the treatment volume can be regulated to meet peak flow requirements (CIRIA, 2000a,b). ♦ An overflow should be provided to deal with large storms and to ensure that a minimum freeboard is maintained in the basin (CIRIA, 2000a,b)
Inlet Infrastructure	♦ Energy dissipation and protection should be provided at the inlet to reduce erosion.
Pretreatment	♦ If excessive sediment loads from runoff upstream have not already been removed before runoff enters the basin, the basin should include a sediment forebay to provide the opportunity for larger particles to settle out. ♦ The forebay volume should be about 10% of the water quality volume and should be provided with a fixed vertical sediment depth marker to measure sediment accumulation. ♦ An inlet sediment forebay could consist of a separate basin, or could be formed by building an earthen berm, stone/rock-filled gabion or riprap wall across the upstream portion of the main basin. The plan area should be between 10% and 25% of the total basin area (CIRIA, 2000a,b). ♦ Flow from the inlet basin to the main basins should be arranged so that settled solids are not resuspended and floating litter washed downstream (CIRIA, 2000a,b)
Subsoil	♦ The soils used to finish the side slopes and base of the basin need to be suitably fertile, porous and of sufficient depth to ensure healthy vegetation growth (CIRIA, 2000a,b).
Water Table	♦ The base of the basin should not intersect any extreme groundwater table, either at the time of design or at any point in the future. ♦ Where infiltration must be controlled, a liner or membrane should be used.
Micropool Characteristics	♦ If a micropool is included in the design, it should be able to store 15%–25% of the capture volume. The larger end of this range is generally preferred to prevent the micropool from drying out during drought periods.

Design Criteria	Current Guidance/Recommendations
Basin Sizing	♦ To enhance the effectiveness of BMP basins, the dimensions of the basin must be sized appropriately. Merely providing the required storage volume will not ensure maximum constituent removal. By effectively configuring the basin, the designer will create a long flow path, promote the establishment of low velocities, and avoid having stagnant areas of the basin. ♦ To promote settling and to attain an appealing environment, the design of BMP basin should consider the length to width ratio, cross-sectional areas, basin slopes and pond configuration, and aesthetics (Young et al., 1996). ♦ A high aspect ratio improves the performance of detention basins; consequently, the outlets should be placed to maximize the flowpath through the facility. The ratio of flow-path length to width from the inlet to the outlet should be at least 1.5:1 (length/width).
Vehicular Access	♦ There should be adequate vehicular access to the main basin, inlet and outlet structures, settling pond and the dry weather channel, so that sediment can be removed during maintenance, and to allow grass cutting and vegetation control.

A variation of the dry detention pond design is the use of tank storage. In these designs, stormwater runoff is conveyed to large storage tanks or vaults underground. This practice is most often used in the ultra-urban environment, on small sites where no other opportunity is available to provide flood control. Tank storage is provided on small areas because providing underground storage for a large drainage area would generally be cost-prohibitive. Because the drainage area contributing to tank storage is typically small, the required outlet diameter will also be very small creating high clogging potential. Since it is necessary to control small runoff events (such as the runoff from a 1-in. storm) to improve water quality, it is generally infeasible to use tank storage for water quality and generally impractical to use it to protect stream channels.

In cold climates, some additional design features can help to treat the spring snowmelt. One such modification is to increase the volume available for detention to help treat this relatively large runoff event. In some cases, dry facilities may be an option as a snow storage facility to promote some treatment of plowed snow. If a pond is used to treat road runoff or is used for snow storage, landscaping should incorporate salt-tolerant species. Finally, sediment might need to be removed from the forebay more frequently than in warmer climates (see Maintenance Considerations guidelines) to account for sediment deposited as a result of road sanding.

Detention basins are normally dry, but safety should be considered for both wet and dry conditions. Potential problems arising during the filling of the pond should be assessed. Shallow side slopes and easy access enable people to leave the area as it slowly floods. A safety review should be carried out and, if necessary, a barrier provided. Barrier planting is a preferable alternative to conventional fencing that provides greater value in terms of visual amenity and provision of wildlife habitat. It is used extensively in Scandinavia (CIRIA, 2000a,b).

2.6 Construction Guidelines

- The bottom and side slopes of the basin should be carefully prepared to ensure that they are structurally sound, level, and true. The preparation should also ensure that the basin will satisfactorily retain the surface water runoff without significant erosion damage (CIRIA, 2000a,b).

- The phasing of the construction should take into account the drainage needs during building and the subsequent performance of the basin (CIRIA, 2000a,b).
- If possible, runoff (other than necessary irrigation) should be diverted during the period of vegetation establishment. Where runoff diversion is not possible, graded and seeded areas should be covered with suitable erosion control materials.

2.7 Performance

One objective of stormwater management practices can be reduction of the flood hazard associated with large storm events by reducing the peak flow associated with these storms. Dry extended detention basins can easily be designed for flood control, and this is actually the primary purpose of most extended detention ponds.

One result of urbanization is the geomorphic changes that occur in response to modified hydrology. Traditionally, dry extended detention basins have provided control of the two-year storm (i.e., the storm that occurs, on average, once every two years) for channel protection. It appears that this control has been relatively ineffective, and recent research suggests that control of a smaller storm might be more appropriate (MacRae, 1996). Slightly modifying the design of dry extended detention basins to reduce the flow from smaller storm events might make them effective in reducing downstream erosion.

Dry extended detention basins provide moderate pollutant removal, provided that the design features described in the Siting and Design Considerations section are incorporated. The smaller storm not only contributes to downstream erosion, but also carries the bulk of the pollutant load; therefore, increasing retention time of these smaller events will also facilitate water quality enhancement. Although they can be effective at removing some pollutants through settling, they are less effective at removing soluble pollutants because of the absence of a large permanent pool.

A few studies are available on the effectiveness of dry extended detention ponds. The California Department of Transportation has conducted one of the more thorough evaluations of extended detention basin performance. Removal rates of the four earthen basins are shown in Table 2-2. These basins were all designed to drain from the full condition in 72 hours and had a minimum length to width ratio of 3:1. The concentrations are the mean of the event mean concentrations (EMCs) for the entire monitoring period. The column titled "Significance" indicates the probability that the influent and effluent concentrations are not significantly different, based on an analysis of variance (ANOVA).

Table 2-2. Pollutant Removal for Extended Detention Basins (Caltrans, 2002).

Constituent	Mean EMC Influent (mg/L)	Mean EMC Effluent (mg/L)	Removal %	Significance *P*
TSS	137	39	72	<.000
NO_3-N	1.06	0.98	8	.529
TKN	2.24	1.85	17	.206
Total N[a]	3.30	2.83	14	NA
Phosphorus	0.52	0.32	39	.001
Orthophosphate	0.11	0.14	-22	.332
Particulate P	0.52	0.32	39	<.000
Total Cu	0.053	0.022	58	<.000
Total Pb	0.087	0.024	72	<.000
Total Zn	0.418	0.115	73	<.000
Dissolved Cu	0.012	0.012	0	.899
Dissolved Pb	0.003	0.002	29	.078
Dissolved Zn	0.071	0.060	16	.279
Particulate Cu	0.041	0.010	76	<.000
Particulate Pb	0.084	0.022	74	<.000
Particulate Zn	0.347	0.055	84	<.000
TPH-Oil [c]	2.80	2.30	18	.773
TPH-Gasoline [c]	0.050[b]	0.050[b]	NA	.974
TPH-Diesel [c]	1.90	1.30	32	.321
Fecal Coliform [c]	900 MPN/100 mL	2000 MPN/100 mL	-122	.607

[a] Sum of NO_3-N and TKN-N
[b] Equals value of reporting limit
[c] TPH and coliform are collected by grab method and may not accurately reflect removal

Schueler (1997a) also reported on the average effectiveness of a number of detention basins (Table 2-3). Removals are generally better in the Caltrans study except for total nitrogen and bacteria, likely because the detention times were longer than those in most other studies.

Table 2-3. Percent Removal for Extended Detention (ED) Basins (Schueler, 1997a).

	TSS	TP	TN	NO_x	HC	Bacteria	Cd	Cu	Pb	Zn
Dry ED Pond (6 studies)	61	20	31	-2	0	78	32	26	54	26

One basin in the Caltrans study was concrete lined and often had higher output concentrations than input. This illustrates the concern that a detention basin, if poorly designed, can act as a source of pollutants, transporting or resuspending material, deposited during previous events (Pratt, 2001). Conversely, good design and operation practices may promote higher performance levels than previously observed.

2.8 Maintenance

Routine maintenance is often thought to consist mostly of sediment and trash and debris removal; however, these activities often constitute only a small fraction of the maintenance hours.

Figure 2-3 presents the average annual maintenance hours experienced in a recent study by the California Department of Transportation.

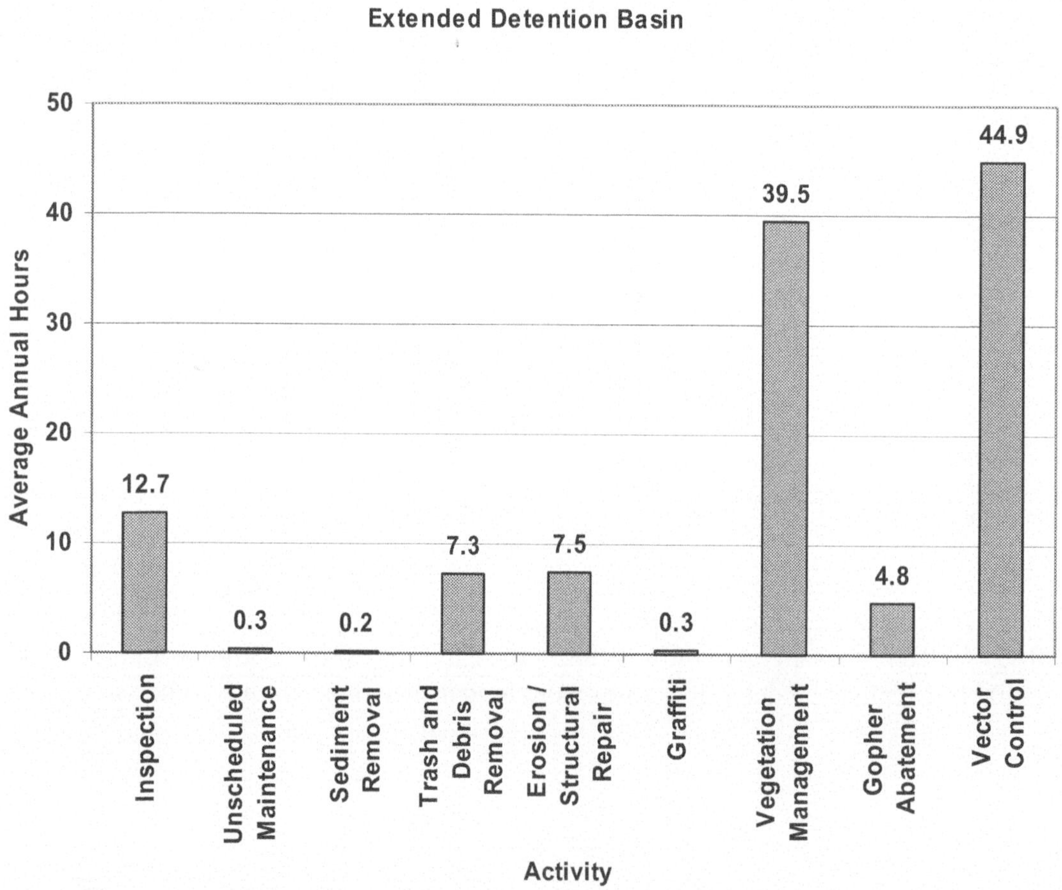

Figure 2-3. Maintenance Activities at Extended Detention Basins (1999–2001).

Excluding vector control, 72 hours of maintenance was performed annually, and approximately seven hours were spent on sediment and trash removal. The largest recurring activity was vegetation management and routine mowing. The largest absolute number of hours was associated with vector control because of mosquito breeding that occurred in the stilling basins installed as energy dissipaters.

In addition to incorporating features into the pond design to minimize maintenance, some regular maintenance and inspection practices are needed. Table 2-4 outlines some of these practices.

Table 2-4. Typical Maintenance Activities for Dry Ponds.

(Source: Modified from WMI, 1997)

Activity	Schedule
♦ Note erosion of pond banks or bottom. ♦ Repair undercut or eroded areas.	Biannually
♦ Inspect for damage to the embankment. ♦ Monitor for sediment accumulation in the facility and forebay. ♦ Examine to ensure that inlet and outlet devices are free of debris and operational.	Quarterly for first year, then annually.
♦ Mow side slopes. ♦ Manage pesticide and nutrients. ♦ Inspect inlet and outlet structures. ♦ Remove litter and debris.	Standard maintenance (at least biannually and after large storms).
♦ Seed or sod to restore dead or damaged ground cover.	Annually
♦ Remove sediment from the forebay.	Post-construction, then 5- to 7-year maintenance.
♦ Monitor sediment accumulations, and remove sediment when the pond volume has been reduced by 25%.	25- to 50-year maintenance

2.9 Design Life

- Design life will depend on design and on effectiveness of ongoing operation and maintenance activities.
- 20 to 50 years' effective life (FWHA, 1999).

2.10 Costs

2.10.1 Construction Costs

Dry extended detention ponds are the least expensive stormwater management practice, on the basis of cost per unit area treated. The construction costs associated with these facilities range considerably. One recent study evaluated the cost of all pond systems (Brown and Schueler, 1997). Adjusting for inflation, the cost of dry extended detention ponds can be estimated with the equation:

$$C = 12.4V^{0.760}$$

where:

C = Construction, design, and permitting cost
V = Volume (ft^3)

Using this equation, typical construction costs are:

\$41,600 for a 1200 m^3 (1 ac-ft) pond
\$239,000 for a 12,000 m^3 (10 ac-ft) pond
\$1,380,000 for a 120,000 m^3 (100 ac-ft) pond

Interestingly, these costs are generally slightly higher than the cost of wet ponds on a cost per total volume basis, which highlights the difficulty of developing reasonably accurate construction estimates. Clearly, dry extended detention ponds are generally less expensive on a given site because they are usually smaller than a wet pond designed for the same site.

Table 2-5 presents the construction costs of a number of facilities located across the US compiled by Caltrans (2001) and ranked by normalized cost. It is very apparent from this list that the cost can vary dramatically, so it would not be justified to use the average to estimate the construction cost for a specific site.

There are several detention basin studies in Dunfermline, Scotland (Spitzer, October 2000–ongoing, in press; Jefferies, 2001); however, studies of extended detention basins are unavailable, most probably due to the fact that their use in the United Kingdom is not widespread. Consequently, there are currently no capital costs available for United Kingdom installations.

An economic concern associated with dry ponds is that they might detract slightly from the value of adjacent properties. One study found that dry ponds can actually detract from the perceived value of homes adjacent to a dry pond by between 3% and 10% (Emmerling-Dinovo, 1995).

Table 2-5. Detention Basin Construction Costs (US\$, adjusted to the Los Angeles region, 1999).

Entity	Drainage Area (acres)	Water Quality Volume (ft^3)	Adjusted Total Cost	Adjusted Total Cost per Acre Treated
Maryland & Virginia CWP	35.00	16,155	\$ 12,558	\$ 359
Maryland & Virginia CWP	77.00	36,435	\$ 34,882	\$ 448
Maryland & Virginia CWP	380.80	1,434,431	\$ 382,556	\$ 1005
Maryland & Virginia CWP	201.00	483,516	\$ 314,033	\$ 1562
MD SHA	145.31	253,200	\$ 267,680	\$ 1842
Maryland & Virginia CWP	3.10	5663	\$ 6813	\$ 2198
Maryland & Virginia CWP	2.30	7841	\$ 6813	\$ 2962
Maryland & Virginia CWP	25.00	222,447	\$ 78,536	\$ 3141
Maryland & Virginia CWP	59.10	655,523	\$ 210,793	\$ 3567
Maryland & Virginia CWP	4.50	9200	\$ 16,464	\$ 3659
Maryland & Virginia CWP	10.90	36,590	\$ 40,752	\$ 3739
Maryland & Virginia CWP	44.30	187,308	\$ 165,652	\$ 3739

Maryland & Virginia CWP	4.30	11,258	$	17,507	$	4071
Maryland & Virginia CWP	19.50	100,188	$	83,875	$	4301
Maryland & Virginia CWP	3.10	10,600	$	13,626	$	4396
Maryland & Virginia CWP	229.90	3,571,920	$	1,041,912	$	4532
Maryland & Virginia CWP	3.20	28,314	$	19,091	$	5966
MD SHA	6.60	3000	$	39,388	$	5968
Maryland & Virginia CWP	3.70	22,651	$	29,319	$	7924
City of Austin	130.00	-	$	1,282,221	$	9863
MD SHA	15.41	61,500	$	155,706	$	10,104
Maryland & Virginia CWP	16.50	222,156	$	206,571	$	12,519
Caltrans	6.80	13,068	$	127,202	$	18,706
ODOT	2.90	-	$	71,540	$	24,669
Caltrans	5.30	13,939	$	147,595	$	27,848
Caltrans	13.40	39,640	$	842,925	$	62,905
Caltrans	4.80	8712	$	339,116	$	70,649
Caltrans	0.80	2614	$	77,389	$	96,737

2.10.2 Maintenance Costs

For ponds, the annual cost of routine maintenance is typically estimated at about 3% to 5% of the construction cost. Alternatively, a community can estimate the cost of the maintenance activities outlined in the maintenance section. Table 2-6 presents the maintenance costs estimated by Caltrans based on its experience with five basins located in southern California. Consistent with the maintenance activities shown in Figure 2-3, the majority of the work consists of vegetation management that may have little impact on basin functionality and is performed primarily for aesthetic reasons.

Table 2-6. Estimated Average Annual Maintenance Effort.

Activity	Labor Hours	Equipment & Material (US$)	Cost (US$)
Inspections	4	7	183
Maintenance	49	126	2282
Vector Control	0	0	0
Administration	3	0	132
Materials	-	535	535
Total	**56**	**$668**	**$3132**

2.11 Research Needs

2.11.1 Siting Criteria

Siting criteria for extended detention basins are well established and no additional work in this area is recommended.

2.11.2 Design Guidelines

There are a number of uncertainties in the specific effect of several commonly adopted design guidelines, including detention time, length to width ratio, and optimum depth. Clearly, a longer detention time allows more of the finer particles to settle out; however, there is a maximum practical limit based on the need to prevent mosquito breeding (72 hours) and to have available storage for subsequent storms. A related issue is a hydrological analysis to assess the frequency of complete filling and how climate change might affect that frequency. Basin geometry and depth are likely of secondary importance in regard to performance (±10 %); consequently, assessment of the effect of these design variables is of less importance. A variety of outlets are recommended for detention basins and additional information to identify the preferred configuration which minimizes maintenance would be useful. Finally, some guidance on integration of these facilities with adjacent areas would provide useful information to project designers.

2.11.3 Performance

There have been numerous studies of pollutant reduction in extended detention basins, so further water quality monitoring is a low priority.

2.11.4 Construction Guidelines and Cost

The wide range of cost data reported indicates that much more information is needed in this area. Do the differences in cost reflect different regulatory requirements concerning inlet and outlet structures, differences in construction techniques, or perhaps site suitability? Identifying the primary elements that drive construction costs upward is a high priority and could help reduce the initial costs associated with these widely used BMP/SUDS.

2.11.5 Maintenance Guidelines and Cost

Maintenance guidelines are one area in which common sense type recommendations are often derived from a common source that may, in fact, have little empirical basis. Many recommendations seem to derive from aesthetic considerations and may have little impact on performance. The only study for which a detailed accounting of maintenance activities was compiled was conducted in a Mediterranean climate by Caltrans and likely is not representative of the frequency or cost of maintenance in other areas. The development of empirical data related to required maintenance activities and their frequency and cost is a high priority.

2.12 Additional Resources

Center for Watershed Protection (CWP), Environmental Quality Resources, and Loiederman Associates, 1997, *Maryland Stormwater Design Manual*. Draft. Prepared for Maryland Department of the Environment, Baltimore, MD.

Center for Watershed Protection (CWP), 1997, *Stormwater BMP Design Supplement for Cold Climates*. Prepared for U.S. Environmental Protection Agency, Office of Wetlands, Oceans and Watersheds. Washington, D.C.

U.S. Environmental Protection Agency (U.S. EPA), 1993, *Guidance Specifying Management Measures for Sources of Nonpoint Pollution in Coastal Waters*. EPA-840-B-92-002. U.S. Environmental Protection Agency, Office of Water, Washington, D.C.

CHAPTER 3.0

WET PONDS/CONSTRUCTED WETLANDS/RETENTION PONDS

3.1 Description

Wet ponds, also known as stormwater ponds, retention ponds, or wet extended detention ponds, are constructed basins that have a permanent pool of water throughout the year (or at least throughout the wet season). They differ from constructed wetlands primarily in having a greater average depth and having peripheral vegetation rather than complete cover. Wet ponds and constructed wetlands treat incoming stormwater runoff through settlement and biological uptake. The primary pollutant removal mechanism is via settlement of silts and suspended sediments. Pollutant uptake, particularly of nutrients, also occurs to some degree through biological activity in the pond. However, to date, no long-term mass balance studies of nutrient uptake and release have been undertaken, and nutrient removal through unattended pond planting remains an area of debate. Wet ponds are among the most widely used stormwater practices. Although there are several different versions of the wet pond design, the most common modification is the extended detention wet pond, where storage is provided above the permanent pool in order to detain stormwater runoff and promote settling.

Wet basins are designed to support emergent and submerged aquatic vegetation along the shoreline, as well as an active microbial community capable of dissolved constituent consumption. If properly designed and sized, sedimentation processes are also effective for capturing the particulate fraction. For most runoff events, pond inflows replace a portion of the prior water quality volume and are stored and "treated" until displaced by the perennial baseflow or next runoff event.

Wet basin permanent pools are generally designed deep enough to prevent resuspension of sediment particles and are sized corresponding to a design storm runoff volume or a factor of the design treatment volume. A number of criteria are available for determining the permanent pool volume. In the United Kingdom, the objective is to provide a residual retention time of 2 to 3 weeks during the wettest months. As well as allowing time for biological treatment to occur, this will also improve the degree of suspended particulate settlement and ultimate removal (CIRIA, 2000a,b).

Enhanced wet basins can be designed with a forebay where trapped coarse sediment can be easily removed during maintenance activities and with aquatic benches that support a fringe wetland along the shore. Shallow depths along the pond perimeter also mitigate the potential hazard of children running into the pond and drowning. Additional benefits include creation of aquatic, wetland, and terrestrial habitats, and high community acceptance, if designed as an attractive urban amenity. A picture of a wet pond is presented in Figure 3-1 and Figure 3-2 shows a schematic representation of a typical wet basin design.

Figure 3-1. Wet Pond, DEX Site, Scotland.

3.2 Advantages

- If properly designed, constructed and maintained, wet basins can provide substantial aesthetic/recreational value as well as wetlands and wildlife habitat.
- Ponds are often viewed as a public amenity when integrated into a park setting.
- Because of the permanent wet pool, properly designed and maintained wet basins can provide significant water quality improvement across a relatively broad spectrum of constituents including dissolved nutrients.
- Widespread use of wet ponds with sufficient capture volume can provide significant control of channel erosion and enlargement caused by changes to flow frequency relationships resulting from the increase of impervious cover in a watershed.

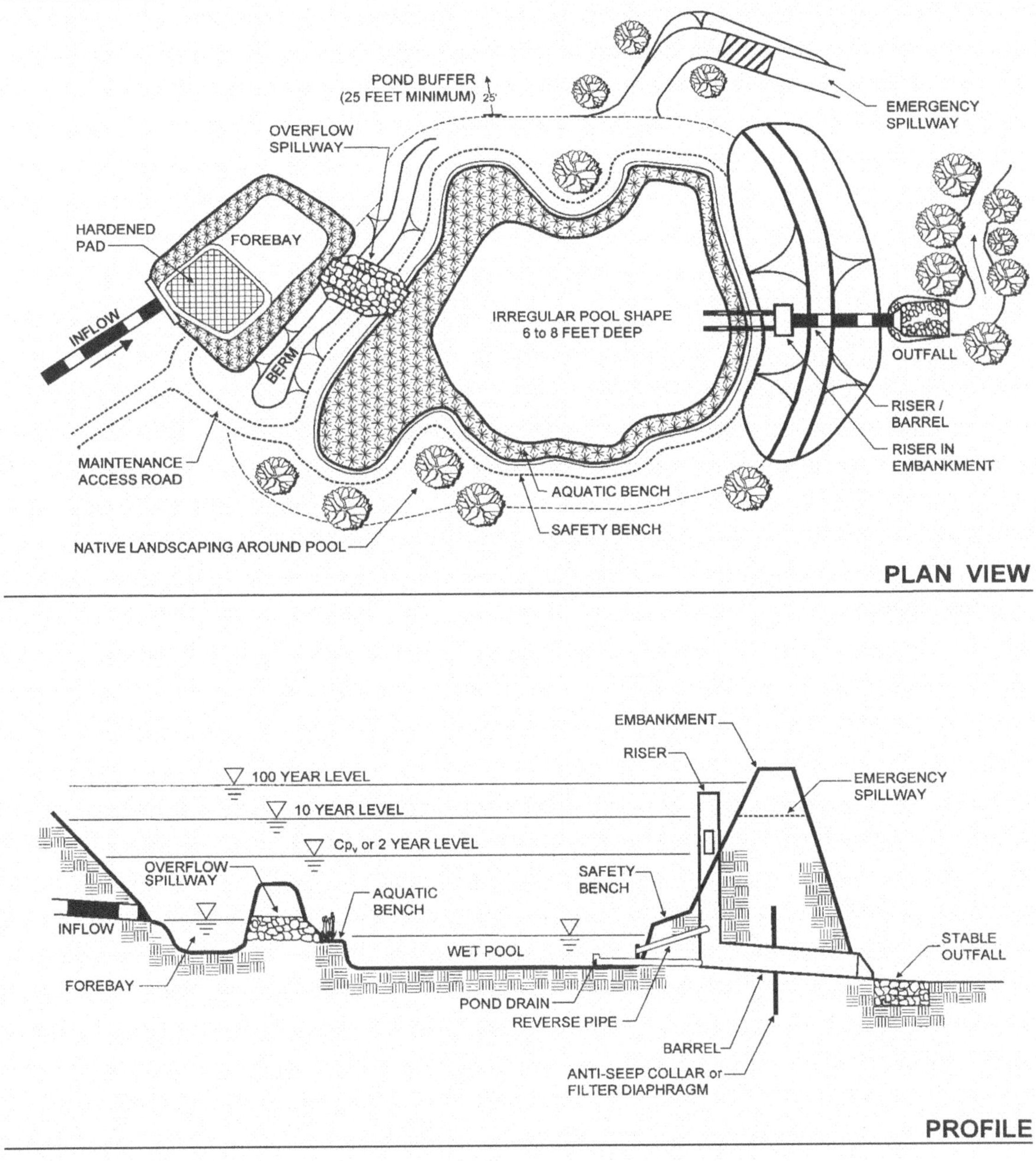

Figure 3-2. Schematic of Wet Basin (Maryland Department of the Environment, 2000).

3.3 Limitations

Wet ponds are widely applicable stormwater management practices. Although they have limited applicability in highly urbanized settings and in arid climates, they have few other restrictions.

- It may be difficult to use retention ponds within densely urbanized areas because of the land required. However, if space is not available on the site itself, consideration could be given to land immediately downstream.

- In locations where clear water is desired for aesthetic reasons, the value of wet ponds for treatment of nutrient rich runoff is limited, as they may suffer from undesirable algae growth during the summer months.
- Liners may be required to sustain a permanent pool all year round, when wet ponds are sited above permeable soils.
- Where groundwater is vulnerable to pollution, a liner must be used to prevent interaction of any local aquifer with the polluted runoff.
- Careful consideration must be given to the siting and design of a wet pond in areas where groundwater levels are high. The systems may not be suitable in these cases if there is a risk that removing ground cover may facilitate the passage of groundwater to the surface during very wet/flood periods.
- Careful consideration is required to minimize health and safety concerns, where these systems form an integral part of any residential development.
- Wet ponds generally require a contributing watershed of more than 5 ha (12.4 ac).
- Recent publicity about West Nile Virus has heightened concern about the potential for vector breeding in facilities incorporating a permanent pool of water.

3.4 Siting Criteria

Wet basins may be constructed on- or off-line and can be sited at feasible locations along established drainage paths with consistent baseflow. In wet climates, basins also perform well without a continuous base flow, providing the pond base is relatively impermeable. Wet basins are often utilized in smaller subwatersheds and are particularly appropriate in areas with residential land uses or other uses where high nutrient loads are considered to be potential problems (e.g., golf courses). Wet basins are typically used in drainage basins of more than 10 acres and less than 1 mi^2 (Schueler, Kumble, and Heraty, 1992). Emphasis can be placed on siting wet basins in areas where the pond can also function as an aesthetic amenity or in conjunction with other stormwater management functions. Wet basins can be used over a broad range of storm frequencies and sizes, drainage areas, and land use types.

Use of wet basins is appropriate in the following settings:

- Where there is a need to achieve a reasonable high level of dissolved contaminant removal and/or sediment capture;
- In small to medium-sized regional tributary areas with available open space and drainage areas greater than about 10 ha (25 ac);
- Where baseflow rates or other channel flow sources are relatively consistent year-round; and
- In residential settings where aesthetic and wildlife habitat benefits can be appreciated and maintenance activities are likely to be consistently undertaken.

Traditional wet extended detention ponds can be used across the United Kingdom and in most regions of the United States, with the exception of arid climates. In arid regions, it is difficult to justify the supplemental water needed to maintain a permanent pool because of the scarcity of water. Even in semiarid Austin, TX, one study found that 3200 m^3 (2.6 ac-ft) per year of supplemental water was needed to maintain a permanent pool of only 350 m^3 (0.29 ac-ft)

(Saunders and Gilroy, 1997). Seasonal wet ponds (i.e., ponds that maintain a permanent pool only during the wet season) may prove effective in areas with distinct wet and dry seasons; however, this configuration has not been extensively evaluated.

Wet ponds may pose a risk to cold water systems because of their potential for stream warming. When water remains in the permanent pool, it is heated by the sun. A study in Prince George's County, Maryland, found that stormwater wet ponds heat stormwater by about 9°F from the inlet to the outlet (Galli, 1990a).

Unless they receive heavily polluted runoff, ponds in the United States may intersect the ground water table. However, in a number of European countries, such as Germany and the United Kingdom, it is not permitted to infiltrate urban runoff into the ground water table. Some research suggests that pollutant removal is reduced when ground water contributes substantially to the pool volume (Schueler, 1997b); however, this may be the result of groundwater adding substantial amounts of dissolved solids to the runoff.

3.5 Design Guidelines

Specific designs may vary considerably, depending on site constraints or preferences of the designer or community. There are several variations of the wet pond design, including constructed wetlands, and the wet extended detention ponds. Some of these design alternatives are intended to make the practice adaptable to various sites and to account for regional constraints and opportunities. In conventional wet ponds, the open water area should comprise 50% to 75 % or more of the total surface area of the pond. The permanent pool should be no deeper than 2.5 m (8 ft) and should average 1.2–2 m (4–6 ft) deep. The greater depth of this configuration helps limit the extent of the vegetation to the perimeter of the pond. The depth limit prevents anaerobic conditions from developing.

Constructed wetlands generally feature relatively uniformly vegetated areas with depths of one foot or less and open water areas (25% to 50% of the total area) no more than about 1.2 m (4 ft) deep, although design configuration options are relatively flexible. Wetland vegetation is generally composed of diverse local aquatic plant species. Constructed wetlands can be designed online or offline and generally serve smaller drainage areas than wet ponds.

The wet extended detention pond combines the treatment concepts of the dry extended detention pond and the wet pond. In this design, the water quality volume is split between the permanent pool and detention storage provided above the permanent pool. During storm events, water is detained above the permanent pool and released over 24-48 hours. Although not well documented, this design appears to have removal efficiency similar to that of a traditional wet pond, and it consumes less space. Wet extended detention ponds should be designed to maintain at least half the treatment volume of the permanent pool. In addition, designers need to carefully select vegetation to be planted in the extended detention zone to ensure that the selected vegetation can withstand both wet and dry periods.

There are some features, however, that should be incorporated into most wet pond designs. These design features can be divided into five basic categories: pretreatment, treatment, conveyance, maintenance reduction, and landscaping.

Table 3-1. Design Guidelines.

Design Criteria	Current Guidance/Recommendations
Inlet Infrastructure	♦ The inlet should be designed to minimize the velocity of runoff entering the system. Energy dissipation and protection should be provided. ♦ The inlet should be sited to reduce the risk of blockage by vegetation or sediment build-up. ♦ The inlet should be sited as far from the outlet as possible.
Pretreatment	♦ Pretreatment incorporates design features that help to settle out coarse sediment particles. By removing these particles from runoff before they reach the large permanent pool, the maintenance burden of the pond is reduced. ♦ In ponds, pretreatment can be achieved with a sediment forebay. A sediment forebay is a small pool (typically about 10% of the volume of the permanent pool). Coarse particles remain trapped in the forebay, and maintenance is performed on this smaller pool, eliminating the need to dredge the entire pond. ♦ In European settings, sediment forebays are not a desired solution, as they are often considered as untidy. Here, sedimentation forebays that are incorporated into the pond and have sediment traps at their inlet are the preferred solution. ♦ Design features should also be incorporated to ease maintenance of both the forebay and the main pool of ponds. Ponds should be designed with a maintenance access to the forebay to ease this relatively routine ♦ (5–7 year) maintenance activity.
Treatment	♦ Treatment design features help enhance the ability of a stormwater management practice to remove pollutants. The purpose of most of these features is to increase the amount of time that stormwater remains in the pond. ♦ One technique of increasing the pollutant removal efficiency of a pond is to increase the volume of the permanent pool. There are a variety of sizing criteria for determining the volume of the permanent pool, mostly related to the water quality volume (i.e., the volume of water treated for pollutant removal) or the average storm size in a particular area. Designers may consider using a larger volume to meet specific watershed objectives, such as phosphorous removal in a lake system. ♦ Other design features do not increase the volume of a pond, but can increase the amount of time stormwater remains in practice, and eliminate short-circuiting. Ponds should always be designed with a length-to-width ratio of at least 1.5:1, although 5:1 has been cited as preferable (Horner et al., 1994) or 3:1 (CIRIA, 2000a,b). In addition, the design could incorporate features to lengthen the flow path through the pond, such as underwater berms/baffles designed to create a longer route through the pond. Windbreaks may also be needed to reduce the influence of the wind on flow patterns. ♦ Combining these two measures helps ensure that the entire pond volume is used to treat stormwater. Another feature that can improve treatment is to use multiple ponds in series as part of a "treatment train" approach to pollutant removal. This redundant treatment can also help slow the rate of flow through the system. ♦ Ponds should generally have a drain to draw down the pond for the more infrequent maintenance activity of dredging the main cell of the pond. Consideration must be given to managing the drainage during these periods of essential maintenance. Areas away from residential sites may need to be set aside for dewatering of the pond sediments prior to disposal offsite.
Flood Control	♦ The additional detention storage provided for flood control should be limited to 2 m above the permanent pool level to avoid damage to the vegetation in the pond (CIRIA, 2000a,b). ♦ An overflow should be provided to deal with large storms and to ensure that a minimum freeboard is maintained in the basin (CIRIA, 2000a,b).
Pond Side Slopes	♦ Maximum pond side slopes should be limited to one in four to meet safety and maintenance requirements (CIRIA, 2000a,b). ♦ Slope protection may be needed during construction and operation of the basin (CIRIA, 2000a,b).

Design Criteria	Current Guidance/Recommendations
	♦ A vegetated buffer should be preserved around the pond to protect the banks from erosion and provide some pollutant removal before runoff enters the pond by overland flow. ♦ Ponds should also incorporate an aquatic bench (i.e., a shallow shelf with wetland plants) about 3 m wide around the edge of the pond. This feature may provide some pollutant uptake, and it also helps to stabilize the soil at the edge of the pond and enhance habitat and aesthetic value. ♦ Selection of appropriate barrier vegetation can also help discourage toddlers and young children from approaching the pond and can therefore act as an additional safety barrier.
Outlet Infrastructure	♦ The outlet works should provide for a smooth stage-discharge relationship for the discharge of storm flows from the pond. This may be achieved, for example, with a V-notch weir transitioning into a sharp crested horizontal weir. Details on the design of different inlet and outlet control structures are provided in Design of Flood Storage Reservoirs (CIRIA, 1993). By using a multistage outlet, flows in excess of the treatment volume can be regulated to meet peak flow requirements (CIRIA, 2000a,b). ♦ One potential maintenance concern in wet ponds is clogging of the outlet. Ponds should be designed with a nonclogging outlet such as a reverse-slope pipe, or a weir outlet with a trash rack. A reverse-slope pipe draws from below the permanent pool extending in a reverse angle up to the riser and establishes the water elevation of the permanent pool. Because these outlets draw water from below the top level of the permanent pool, they are less likely to be clogged by floating debris. ♦ The outlet should be sited as far from the inlet as possible.

3.5.1 Regional Adaptations

3.5.1.1 Semiarid Climates

In arid climates, wet ponds are not a feasible option (see Applicability), but they may possibly be used in semiarid climates if the permanent pool is maintained with a supplemental water source, or if the pool is allowed to vary seasonally. This choice needs to be seriously evaluated, however.

3.5.1.2 Cold Climates

Cold climates present many challenges to designers of wet ponds. The spring snowmelt may have a high pollutant load and a large volume to be treated. In addition, cold winters may cause freezing of the permanent pool or freezing at inlets and outlets. Finally, high salt concentrations in runoff resulting from road salting, and sediment loads from road sanding, may affect pond vegetation and reduce the storage and treatment capacity of the pond.

One option to deal with high pollutant loads and runoff volumes during the spring snowmelt is the use of a seasonally operated pond to capture snowmelt during the winter, and retain the permanent pool during warmer seasons. In this option, proposed by Oberts (1994), the pond has two water quality outlets, both equipped with gate valves. In the summer, the lower outlet is closed. During the fall and throughout the winter, the lower outlet is opened to draw down the permanent pool. As the spring melt begins, the lower outlet is closed to provide detention for the melt event. This method can act as a substitute for using a minimum extended detention storage volume. When wetlands preservation is a downstream objective, seasonal manipulation of pond levels may not be desired. An analysis of the effects on downstream hydrology should be conducted before considering this option. In addition, the manipulation of this system requires some labor and vigilance; a careful maintenance agreement should be confirmed.

Several other modifications may help to improve the performance of ponds in cold climates. Designers should consider planting the pond with salt-tolerant vegetation if the facility receives road runoff. In order to counteract the effects of freezing on inlet and outlet structures, the use of inlet and outlet structures that are resistant to frost, including weirs and larger diameter pipes, may be useful. Designing structures on-line, with a continuous flow of water through the pond, will also help prevent freezing of these structures. Finally, since freezing of the permanent pool can reduce the effectiveness of pond systems, it may be useful to incorporate extended detention into the design to retain usable treatment area above the permanent pool when it is frozen.

3.6 Construction Guidelines

- The bottom and side slopes of the basin should be carefully prepared to ensure that they are structurally sound, level, and true. The preparation should also ensure that the basin will satisfactorily retain the surface water runoff without significant erosion damage (CIRIA, 2000a,b).
- The phasing of the construction should take into account the drainage needs during building and the subsequent performance of the basin (CIRIA, 2000a,b).
- If possible, divert runoff (other than necessary irrigation) during the period of vegetation establishment. Where runoff diversion is not possible, cover graded and seeded areas with suitable erosion control materials.
- As wet ponds rely to some extent on the establishment of vegetation to enhance the treatment effect, the establishment period should be included within the construction program.
- If an impermeable liner is used, care should be taken to ensure that it is not damaged during construction.

3.7 Performance

Structural stormwater management practices can be used to achieve four broad resource protection goals. These include flood control, channel protection, groundwater recharge, and pollutant removal. Wet ponds can contribute to all of these objectives, except for groundwater recharge.

3.7.1 Hydraulic Performance

One objective of stormwater management practices can be to reduce the flood hazard associated with large storm events by reducing the peak flow associated with these storms. Wet ponds can easily be designed for flood control by providing flood storage above the level of the permanent pool.

When used for channel protection, wet ponds have traditionally provided outlet control for the two-year storm, while allowing smaller storms to pass without attenuation. It appears that this control has been relatively ineffective, and recent research suggests that control of a smaller storm may be more appropriate (MacRae, 1996). There is usually a tradeoff between hydrological control and cost, as control of smaller storms will require more surface area for the pond. It may therefore be necessary to rely on flow control upstream of the pond to achieve more ambitious targets. A close look at the geomorphology of the receiving watercourse is necessary to select the appropriate maximum flow rates at the pond outlet. In general, an outlet design that detains the water quality volume for at least 24 hours is preferred in order to provide some protection of the receiving watercourse.

3.7.2 Pollutant Removal

The observed performance of a wet pond is highly dependent on two factors: the volume of the permanent pool relative to the amount of runoff from the typical event in the area, and the quality of the baseflow that sustains the permanent pool. A recent study (Caltrans, 2002) has documented that if the permanent pool is much larger than the volume of runoff from an average event, then displacement of the permanent pool by the wet weather flow is the primary process. A statistical comparison of the wet pond discharge during dry and wet weather discharge shows that the concentrations are not significantly different. This results in a relatively constant discharge quality during storms that is independent of the influent concentrations. Consequently, for most constituents the performance of the pond is better characterized by the average effluent concentration, rather than the percent reduction, which has been the conventional measure of performance. Since the effluent quality is essentially constant, the percent reduction observed is mainly a function of the influent concentrations observed at this particular site.

The quality of dry weather and wet weather discharges is related to the quality of the baseflow that sustains the permanent pool and of the transformations that occur to those constituents during their residence in the basin. One could potentially expect a wide range of effluent concentrations at different locations even if the wet ponds were designed according to the same guidelines, if the quality of the baseflow differed significantly. This may explain the wide range of concentration reductions reported in the various reports and shown in Table 3-2 and the following graphs.

Median removal rates from the studies listed below are as follows:

Total Suspended Solids:	78%	Nitrate Nitrogen:	32%
Total Phosphorus:	47%	Metals:	60%
Total Nitrogen:	32%	Bacteria:	70%

Table 3-2. Wet Pond Pollutant Removal Efficiencies.

Study	TSS %	TP %	TN %	NO_3 %	Metals %	Bacteria %	Practice Type
City of Austin, TX 1991. Woodhollow, TX	54	46	39	45	69-76	46	wet pond
Driscoll 1983. Westleigh, MD	81	54	37	-	26-82	-	wet pond
Dorman et al., 1989. West Pond, MN	65	25	-	61	44-66	-	wet pond
Driscoll, 1983. Waverly Hills, MI	91	79	62	66	57-95	-	wet pond
Driscoll, 1983. Unqua, NY	60	45	-	-	80	86	wet pond

Study	TSS %	TP %	TN %	NO_3 %	Metals %	Bacteria %	Practice Type
Cullum, 1985. Timber Creek, FL	64	60	15	80	-	-	wet pond
City of Austin, TX 1996a. St. Elmo, TX	92	80	19	-17	2-58	89-91	wet pond
Horner, Guedry, and Kortenhoff, 1990. SR 204, WA	99	91	-	-	88-90	-	wet pond
Horner, Guedry, and Kortenhoff, 1990. Seattle, WA	86.7	78.4	-	-	65-67	-	wet pond
Kantrowitz and Woodham, 1995 Saint Joe's Creek, FL	45	45	-	36	38-82	-	wet pond
Wu, 1989. Runaway Bay, NC	62	36	-	-	32-52	-	wet pond
Driscoll, 1983. Pitt-AA, MI	32	18	-	7	13-62	-	wet pond
Bannerman and Dodds, 1992. Monroe Street, WI	90	65	-	-	65-75	70	wet pond
Horner, Guedry, and Kortenhoff, 1990. Mercer, WA	75	67	-	-	23-51	-	wet pond
Oberts, Wotzka, and Hartsoe, 1989. McKnight, MN	85	48	30	24	67	-	wet pond
Yousef, Wanielista, and Harper 1986. Maitland, FL	-	-	-	87	77-96	-	wet pond
Wu, 1989. Lakeside Pond, NC	93	45	-	-	80-87	-	wet pond
Oberts, Wotzka, and Hartsoe, 1989. Lake Ridge, MN	90	61	41	10	73	-	wet pond
Driscoll, 1983. Lake Ellyn, IL	84	34	-	-	71-78	-	wet pond
Dorman et al., 1989. I-4, FL	54	69	-	97	47-74	-	wet pond
Martin, 1988. Highway Site, FL	83	37	30	28	50-77	-	wet pond
Driscoll, 1983. Grace Street, MI	32	12	6	-1	26	-	wet pond
Occoquan Watershed Monitoring Laboratory, 1983. Farm Pond, VA	85	86	34	-	-	-	wet pond
Occoquan Watershed Monitoring Laboratory, 1983. Burke, VA	-33.3	39	32	-	38-84	-	wet pond
Dorman et al., 1989. Buckland, CT	61	45	-	22	-25 to -51	-	wet pond
Holler, 1989. Boynton Beach Mall, FL	91	76	-	87	-	-	wet pond

Study	TSS %	TP %	TN %	NO_3 %	Metals %	Bacteria %	Practice Type
Urbonas, Carlson, and Vang 1994. Shop Creek, CO	78	49	-12	-85	51-57	-	wet pond
Oberts and Wotzka, 1988. McCarrons, MN	91	78	85	-	90	-	wet pond
Gain, 1996. FL	54	30	16	24	42-73	-	wet pond
Ontario Ministry of the Environment, 1991. Uplands, Ontario	82	69	-	-	-	97	wet extended detention pond
Borden et al., 1996. Piedmont, NC	19.6	36.5	35.1	65.9	-4 to -97	-6	wet extended detention pond
Holler, 1990. Lake Tohopekaliga District, FL	-	85	-	-	-	-	wet extended detention pond
Ontario Ministry of the Environment, 1991. Kennedy-Burnett, Ontario	98	79	54	-	21-39	99	wet extended detention pond
Ontario Ministry of the Environment, 1991. East Barrhaven, Ontario	52	47	-	-	-	56	wet extended detention pond
Borden et al., 1996. Davis, NC	60.4	46.2	16	18.2	15-51	48	wet extended detention pond
Hopwood — Pond 2 (Folkes-Skinner, 2002)	18			30	Cd, Ni: 0; Cr: 15; Zn: 65; Pb: 72; Cu: 32		wet pond
Summary of U.K. data - Urban Runoff (Ellis et al., 2001)	55			29			wet pond
Summary of UK data - Highway Runoff (Ellis et al., 2001)					Zn: 38; Pb: 52		wet pond
Newbury — Pond B2 (Shutes et al., 2001)	-6				Cd: 8, Ni: 52; Cr: 15; Zn: 29; Pb: 0; Cu: -75		wet pond
Claylands Pond (Jefferies, 2001)					Ni: 61; Cr:80 ; Zn:42 ; Pb:75 ; Cu: 77		

Figure 3-3 through 3-6 graphically illustrate wet pond performance for selected constituents. Most of the data indicate substantial and consistent removal of TSS; however, the data for nutrients shows much higher variability. Reasons for this variability were described by O'Shea, Borst, and Nietch (2002) and are reproduced below.

According to O'Shea, Borst, and Nietch (2002), data on phosphorus retention by stormwater BMPs often confusedly reflect a mix of short-term (e.g., plant uptake, exchangeable sorption) and longer-term (e.g., sedimentation, burial) removal processes. Monitoring data collected only during the growing season may overestimate capture since much of the nutrient load incorporated into plant or microbial biomass during growth phases will be released upon die back. Similarly, phosphorus removed via sorption processes can be rapidly released into the water column upon changing (e.g., seasonal) water column or sediment conditions (e.g., redox potential or pH). In addition to seasonal differences, phosphorus removal in newly constructed BMPs can also significantly decline after the first few years of treatment due to the saturation of sorption sites. A variety of soil amendments have been examined as a means to increase the sorption capacity of wetland or filtration media.

O'Shea and colleagues (2002) also reports that in the absence of harvesting measures aimed at removing organically bound nutrients (and generally considered not cost-effective), burial in the sediment layer is considered one of the only processes capable of providing substantial and long-term phosphorus retention. Burial can occur by several means, including the settling of incoming solids and their associated phosphorus load, the formation of chemical precipitates, sorption onto solids or biofilms in the water column or sediment layer, and the accumulation and burial of refractory organic matter. However, not all of these processes can be considered irreversible. For example, settled phosphorus can become resuspended by mixing events, often in a more bioavailable form due to diagenetic transformations. Other processes, such as sorption or precipitation, may provide only temporary detention, ultimately operating in reverse to release phosphorus into surface waters.

Nitrogen, like phosphorus, is present in receiving waters and surface runoff in a number of forms, including dissolved and particulate, organic and inorganic species. Dissolved inorganic nitrogen (DIN) species, in particular ammonium (NH_{4+}), are generally considered the most available for immediate assimilation by aquatic micro- and macrobiota. However, like phosphorus uptake, biotic uptake of nitrogen is generally considered to provide mainly short-term (i.e., seasonal), nitrogen removal, with much of the stored nutrients released as dissolved (DON) and particulate (PON) organic species during fall/winter senescence.

Options for long-term removal of nitrogen include not only burial, as with phosphorus, but also the biologically mediated sequential reactions, ammonification (the mineralization of organic nitrogen into ammonia and ammonium (NH_3/NH_{4+}), nitrification (the oxidation of NH_3/NH^{4+} to nitrate (NO^{3-}) and denitrification (the anaerobic reduction of NO^{3-}) that results in the gaseous end products nitric oxide (NO), nitrogen gas (N_2), and nitrous oxide (N_2O). Together with the pH-mediated formation of volatile NH_3, the formation and export of these gaseous species generally account for a significant portion of the nitrogen loss observed in receiving waters and treatment systems. Treatment systems often attempt to promote these reactions by providing the conditions and residence times necessary for each transformation.

Several factors will affect the extent of ammonification, nitrification, and denitrification. Some of the more important include the form, concentration, and timing of nitrogen loads; the macro- and microbiotic communities; the availability of a carbon supply sufficient to fuel denitrification; and the chemical and physical conditions (e.g., temperature, pH, and dissolved oxygen) within a system. Export, rather than retention, of any single species may occur in the absence of conditions or reactants necessary to maintain reaction rates. For example, wetlands receiving influent high in NH_4^+ or organic nitrogen (rather than NO^{3-})have been observed to export total nitrogen. Similarly, an inadequate or unavailable carbon source in newly created or subsurface flow wetlands may limit denitrification, leading to NO_3^- export.

In contrast to the dissolved inorganic phosphorus (DIP) species, NH_4^+ is the only form of DIN that can be significantly removed via sorption onto soil minerals. As with phosphorus, studies have attempted to increase NH_4^+ removal through soil amendment (Phillips, 1998). However, as in natural systems, removals will be finite unless sorption sites are chemically or biologically regenerated. Nitrate, in contrast to NH_4^+ and DIP, is quite mobile and likely to be physically removed from BMPs only through groundwater flow. For this reason, NO_3^- is a contaminant of concern in many aquifer source waters and may additionally impact receiving waters that receive substantial groundwater inputs.

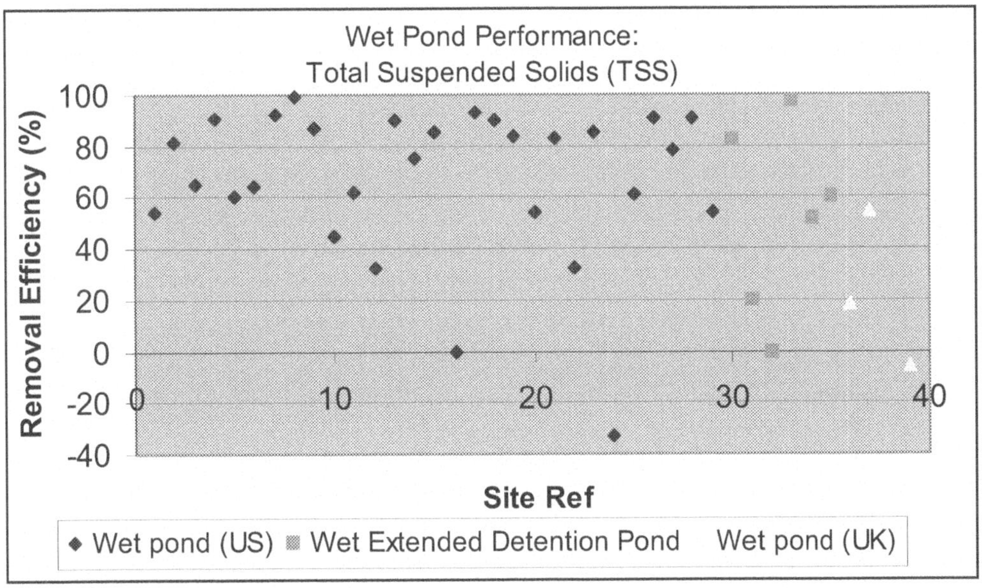

Figure 3-3. TSS Removal in Wet Ponds.

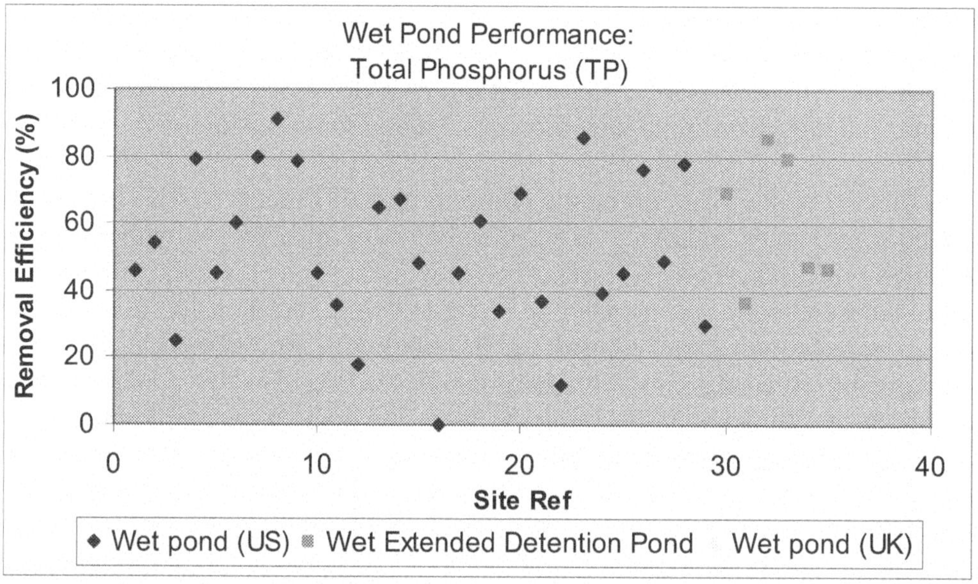

Figure 3-4. Phosphorus Removal in Wet Ponds.

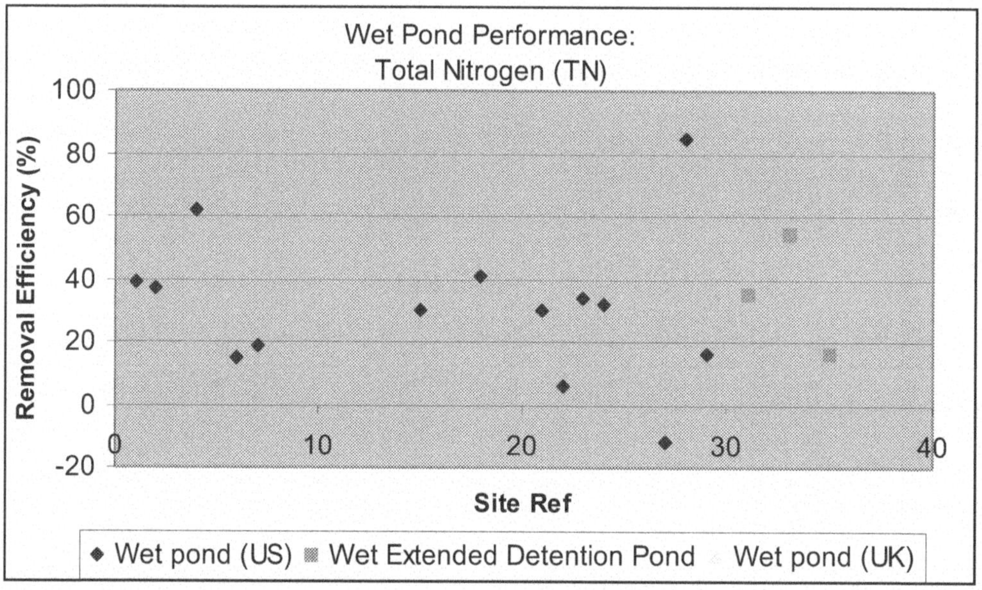

Figure 3-5. Total Nitrogen Removal in Wet Ponds.

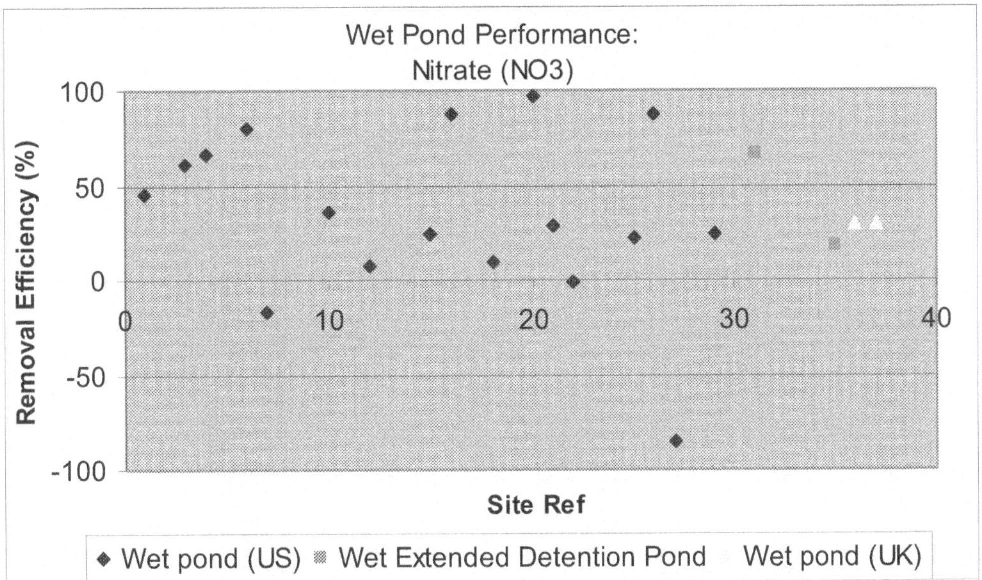

Figure 3-6. Nitrate Removal in Wet Ponds.

3.8 Maintenance

Maintenance requirements of wet ponds vary greatly depending on local regulations, aesthetic considerations, the quality of the baseflow that sustains the permanent pool, and climatic considerations. In Southern California, vector concerns (mosquito breeding, primarily) was one of the major drivers for maintenance. Local vector control agencies required annual removal of the majority of the vegetation to allow access to the mosquito larvae by the mosquito fish (*Gambusia affinis*). This vegetation harvesting accounts for the vast majority of the 250 hours/year estimated for future maintenance activities at a fairly small facility that treated the runoff from only 1.7 ha (4.2 ac) (Caltrans, 2002). The breakdown of hours actually incurred by Caltrans in a very aggressive maintenance program associated with research study is shown in Figure 3-7. The vegetation at this site was particularly vigorous because of the high nutrient concentrations in the perennial baseflow (15.5 mg/L NO_3-N) and the mild climate, which permits growth year round. It should be noted that vegetation harvesting removes nutrients from the system, preventing recycling, and potentially improving the removal of nutrients from stormwater.

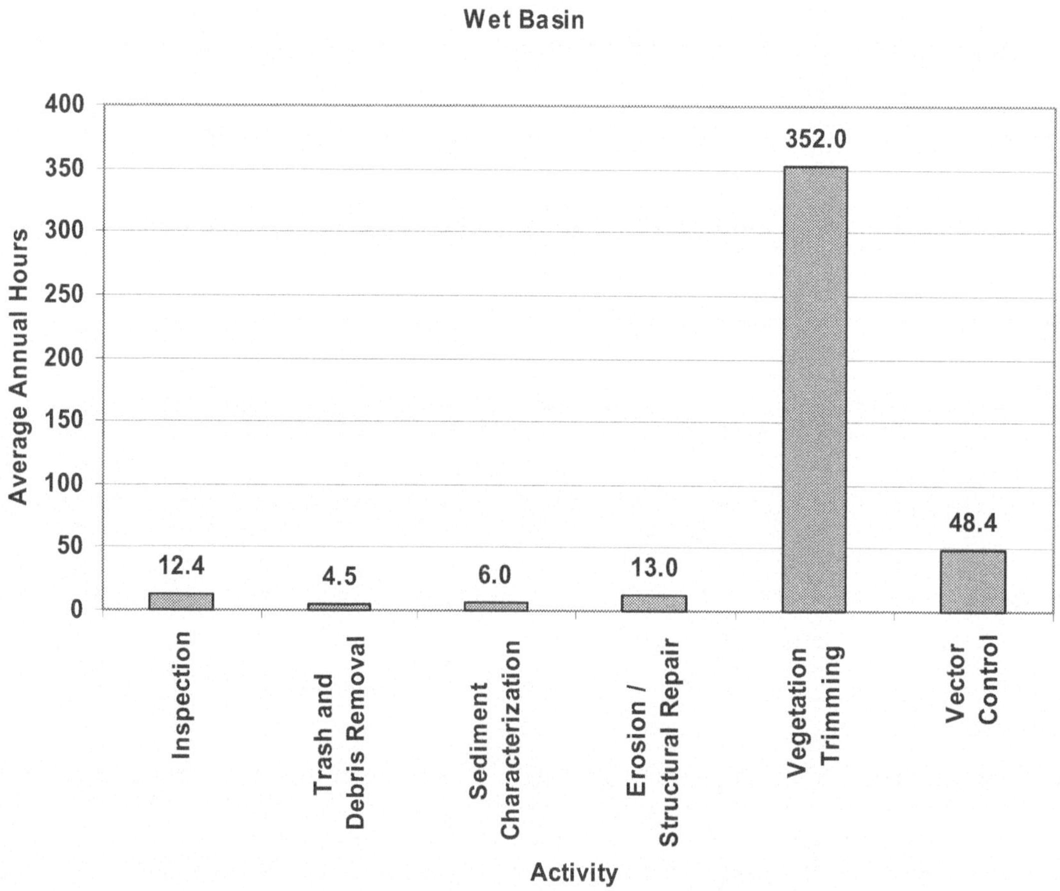

Figure 3-7. Actual Maintenance Hours for a Wet Basin in Southern California.

Because wet ponds are often selected for aesthetic considerations as well as pollutant removal, they are often sited in areas of high visibility. Consequently, floating litter and debris are removed more frequently than would be required simply to support proper functioning of the pond and outlet. This is one of the primary maintenance activities performed at the Central Market Pond located in Austin, Texas (Glick, 2001). In this type of setting, vegetation management in the area surrounding the pond can also contribute substantially to the overall maintenance requirements.

One normally thinks of sediment removal as one of the typical activities performed at stormwater BMPs. This activity does not normally constitute one of the major activities on an annual basis. At the concentrations of TSS observed in urban runoff from stable watersheds, sediment removal may only be required every 20 years or so. Because this activity is performed so infrequently, accurate costs for this activity are lacking.

In addition to incorporating features into the pond design to minimize maintenance, some regular maintenance and inspection practices are needed. Table 3-3 outlines these practices.

Table 3-3. Typical Maintenance Activities for Wet Ponds (Source: WMI, 1997).

Activity	Schedule
♦ If wetland components are included, inspect for invasive vegetation.	Semi-annual inspection
♦ Inspect for damage. ♦ Note signs of hydrocarbon build-up and deal with appropriately. ♦ Monitor for sediment accumulation in the facility and forebay.	Annual inspection
♦ Collect litter.	Monthly
♦ Examine to ensure that inlet and outlet devices are free of debris and operational. ♦ Clean and remove debris from inlet and outlet structures	Quarterly, and following large storms
♦ Repair undercut or eroded areas.	As needed maintenance
♦ Mow side slopes.	Monthly maintenance
♦ Manage and harvest wetland plants.	Annual maintenance (if needed)
♦ Remove sediment from the forebay.	Post construction and then 5-7 year maintenance regime (or once 50% of forebay capacity has been lost)
♦ Maintain fence.	As needed maintenance
♦ Monitor sediment accumulations, and remove sediment when the pool volume has become reduced significantly or the pond becomes eutrophic.	20-50 year maintenance (or when sediment accumulation is significant)

3.9 Design Life

- 20–50 years (FWHA, 1999).
- 20 years or longer if well designed and maintained (U.S. EPA, 1999e).

3.10 Costs

There is little published information on pond maintenance or costs in the UK. Maintenance monitoring of four ponds at Duloch Park, Dunfermline, and one wetland commenced in the spring of 1999 by the University of Abertay Dundee under a contract with Wilcon Home (Scotland). The results are due in mid-2004, following the collection of 5 years of maintenance activity data and cost information.

3.10.1 Construction Costs

Wet ponds can be relatively inexpensive stormwater practices; however, the construction costs associated with these facilities vary considerably. Much of the variability in costs can be attributed to the degree to which the existing topography will support a wet pond, the complexity and amount of concrete required for the outlet structure, and whether it is installed as part of new construction or implemented as a retrofit of an existing storm drain system. Value of land and loss of space for property development may add to the cost of the drainage infrastructure.

A recent study (Brown and Schueler, 1997) estimated the cost of a variety of stormwater management practices. The study resulted in the following cost equation, adjusting for inflation:

$$C = 24.5V^{0.705}$$

where:

C = Construction, design and permitting costs
V = Volume in the pond to include the 10-year storm (ft^3)

Using this equation, typical construction costs are:
\$45,700 for a 1200 m^3 (1 ac-ft) facility
\$232,000 for a 12,000 m^3 (10 ac-ft) facility
\$1,170,000 for a 120,000 m^3 (100 ac-ft) facility

In contrast, Caltrans (2002) reported spending over \$448,000 for a pond with a total permanent pool plus water quality volume of only 1036 m^3 (0.8 ac-ft), while the City of Austin spent \$584,000 (including design) for a pond with a permanent pool volume of 3100 m^3 (2.5 ac-ft).

Ponds do not consume a large area relative to the contributing drainage area (typically 2%–3%); however, these facilities are generally large relative to other BMPs. Other practices, such as filters or swales, may be squeezed into relatively unusable land, but ponds need a relatively large continuous area.

Typical costs for wet ponds range form \$17.50–\$35.00 per cubic meter (\$0.50–\$1.00 per ft^3) (CWP, 1998). Several studies have shown retrofitting a wet pond to a developed area may be 5–10 times the costs of constructing the same size pond in an undeveloped area. (U.S. EPA, 1999e).

No cost information is available from the United Kingdom.

3.10.2 Maintenance Costs

For ponds, the annual cost of routine maintenance has typically been estimated at about 3% to 5% of the construction cost; however, the published literature is almost totally devoid of actual maintenance costs. Since ponds are long-lived facilities (typically longer than 20 years), major maintenance activities are unlikely to occur during a relatively short study.

Table 3-4 presents the annual maintenance costs estimated by Caltrans (2002) based on three years of monitoring of a pond treating runoff from 1.7 ha (4.2 ac). Labor hours were converted to cost assuming a fully burdened rate of \$44/hour. As mentioned previously, almost all the activities under the heading "Maintenance" are associated with vegetation management. Consequently, maintenance costs may vary considerably at other sites depending on the aggressiveness of the vegetation management in that area. Total cost at this site falls within the 3%–5% range reported above; however, the construction costs were much higher than those incurred at other locales.

Table 3-4. Expected Annual Maintenance Costs.

Activity	Labor Hours	Equipment & Material ($)	Cost
Inspections	8	0	352
Maintenance	262	375	11,903
Vector Control	12	0	744
Administration	3	0	133
Materials	-	4500	4500
Total	**285**	**$4875**	**$17,631**

3.11 Research Needs

Research needs relative to wet pond performance are described below.

3.11.1 Siting Criteria

Siting criteria for wet ponds are well established, and no additional work in this area is recommended.

3.11.2 Design Guidelines

The primary uncertainty related to current design guidelines is the size of the permanent pool relative to the design storm one wishes to treat. The size of the permanent pool is often recommended to be a multiple (one to three times) of the average storm for an area. There is little data to substantiate these recommendations and more research is recommended.

Other design issues needing research include:

- Ability to consistently achieve target detention times across all expected storm events; and
- Ability to design for treatment volumes and detention times that best protect downstream channels.

3.11.3 Performance

Much of the uncertainty related to performance is associated with the pollutant removal of the emergent vegetation. Some of the issues include:

- Need for more research into water quality performance, particularly long-term performance with mass balance figures over several years. Some research on water quality performance of Scottish SUDS ponds has been done by Macdonald and Jefferies (2000), McLean (2001), and currently by Spitzer (Schlüter, Spitzer, and Jeffries, 2002). Not enough is known about the effects of the build-up in nutrient levels in vegetation on nutrient through-flow. The benefits of vegetation harvesting, and the best approach, needs to be addressed.
- Uptake of heavy metals.
- Degree to which pollutant removal rates are reduced during nongrowing season.
- More research needed into rate of accumulation of sediments and increasing levels of toxicity. Some research into pollutant accumulation in pond sediments is still ongoing at the University of Edinburgh by Kate Heal.
- Relationship between pond performance and appearance and its perception locally.

3.11.4 Construction Guidelines and Cost

Capital cost will vary widely depending on site and type of pond, but can probably be relatively easily established for each new pond by quantity surveyors. A related issue that needs more research is the influence of the pond on house prices and house salability. In addition research is needed into the effectiveness of different marginal arrangements in providing safety barriers.

3.11.5 Maintenance Guidelines and Cost

Maintenance issues that warrant further research include:

- Data for routine maintenance costs are currently being gathered for four ponds and one wetland at DEX, Dunfermline (along with information on several detention basins and swales). No information about the long-term maintenance of ponds has been found.
- Impact of differing maintenance regimes on pond performance (hydraulics and water quality).
- Real maintenance needs associated with sediment management (as this is currently largely being ignored).

3.12 Additional Resources

Behrendt, L., 2002, Diversity and steadiness of aquatic macrophytes in retention ponds of Central Scotland as part of "Sustainable Urban Drainage Systems" and consideration of heavy metal concentrations in plant tissues and marginal sediment. MSc Thesis, Hochschule Zittau/Goerlitz.

Brix, H., 1997, Do macrophytes play a role in constructed treatment wetlands? *Water Science Technology,* 35 pp 11-17.

Foo, E.K.C., 2000, Coliform removal from urban runoff by passive detention systems. Part of MSc Thesis, University of Edinburgh.

Gumbricht, T., 1993, Nutrient removal processes in freshwater submersed macrophytes systems. *Ecological Engineering, 3*, pp 1-30.

Jefferies, C., Aitken, A., McLean, N. Macdonald, K., McKissock, G., 1999, Assessing the performance of urban BMPs in Scotland. *Water Science Technology, 39*, 12, pp 123-131.

Knight, R.L., 1997, Wildlife habitat and public use benefits of treatment wetlands. *Water Science Technology, 35,* 5, pp 11-17.

Macdonald, K., Jefferies, C., and Dring, D., 1999, An assessment of the efficiency of a stormwater runoff pond in the Scottish climate. IAWQ 8th International Conference on Urban Storm Drainage. 30 August-3 September 1999, Sydney, Australia.

McKissock, G., D'Arcy, B.J., Jefferies, C. 2001. Sustainable urban drainage: a case study. Paper presented on Novatech 2001-Innovative technologies in urban drainage conference proceedings, 25-27 June 2001, Lyon, France. GRAIE, Villeurbanne, Lyon, France.

Pond Action, 2000, *An Assessment of the Ecological Value of Sustainable Urban Drainage System Ponds in Scotland.* Pond Action, Oxford.

Roesner, L.A., Campbell, N., D'Arcy, B.J, 2001, Master planning stormwater management facilities for the Dunfermline, Scotland expansion project. Paper presented at Novatech 2001.

Innovative Technologies in Urban Drainage Conference Proceedings, 25-27 June 2001, Lyon, France. GRAIE, Villeurbanne, Lyon, France.

SEPA, year unknown, *Protecting the Quality: A Guide to Sustainable Urban Drainage.* Scottish Environment Protection Agency, Edinburgh.

SEPA, 2000, *Ponds Pools and Lochans: Guidance on Good Practice in the Management and Creation of Small Water Bodies in Scotland.* Habitat Enhancement Initiative, Scottish Environment Protection Agency, Edinburgh.

Stahre, P., 1999, 10 Years of Experiences of Sustainable Stormwater Management in the City of Malmö, Sweden. 8th International Conference on Urban Storm Drainage 30Aug – 3 Sept. 1999, Proceedings (Volume 3), pp 1087-1097.

Spitzer, A., 2000, Preliminary results from the hydrological and water quality investigations at Duloch Park. Paper presented to SUDS Monitoring Workshop at Aberdeen University on 26 September 2000.

Spitzer, A., 2000, *Hydrological Performance of Two Retention Ponds at Duloch Park. Scottish Universities SUDS Monitoring Programme, Report on First Phase of the Programme 1997 – 2001;* Draft for discussion in workshop on 25 April 2001 at University of Abertay Dundee.

Spitzer, A., 2001, Hydrological investigations at Duloch Park – first results. Paper presented at Postgraduate Student Workshop of University of Abertay Dundee on 21 May 2001.

CHAPTER 4.0

INFILTRATION BASINS

4.1 Description

An infiltration basin (Figure 4-1) is a shallow impoundment which is designed to infiltrate stormwater into the ground. This practice has high pollutant removal efficiency and can also help recharge the groundwater, thus restoring low flows to stream systems. Infiltration basins should only be used as part of a treatment train, wherein soluble organic substances, oils, and course sediment are removed by other management practices prior to stormwater entering the infiltration basin. This practice should not be used in industrial parks, high density or heavy industrial areas, chemical or pesticide storage areas, or fueling stations where there may be a risk to groundwater contamination.

Infiltration basins can be challenging to apply on many sites because of soil characteristics. In addition, some studies have shown relatively high failure rates compared with other management practices as a result of surface clogging and reduction in infiltration potential.

Infiltration basins are more commonly used in North America and most of the information given in this report is based on U.S. experiences. No U.K.-based performance data are available but a number of studies have been undertaken in France and America (CIRIA, 1999). The main difficulty associated with monitoring these systems is that almost all water infiltrates through the basin and quality monitoring would involve assessing the change in soil water/groundwater quality, and there is little data in this area. Infiltration basins have not been employed in Scotland, as ground conditions are generally unsuitable (Pratt, 2001).

4.2 Advantages

- The principal benefit of infiltration basins is the approximation of predevelopment hydrology during which a significant portion of the average annual rainfall runoff is infiltrated and evaporated rather than flushed directly to receiving waters.
- If the infiltration basin volume is adequately sized, the systems can be very useful for providing control of channel forming (erosion) and high frequency (generally less than the 2-year) flood events.
- Although infiltration basins are deemed suitable for treating relatively small catchment areas ranging from 2-6 ha (5-15 acres) (Schueler, 1987), experience at Venissieux and Fresno, California shows that they can serve much larger areas.

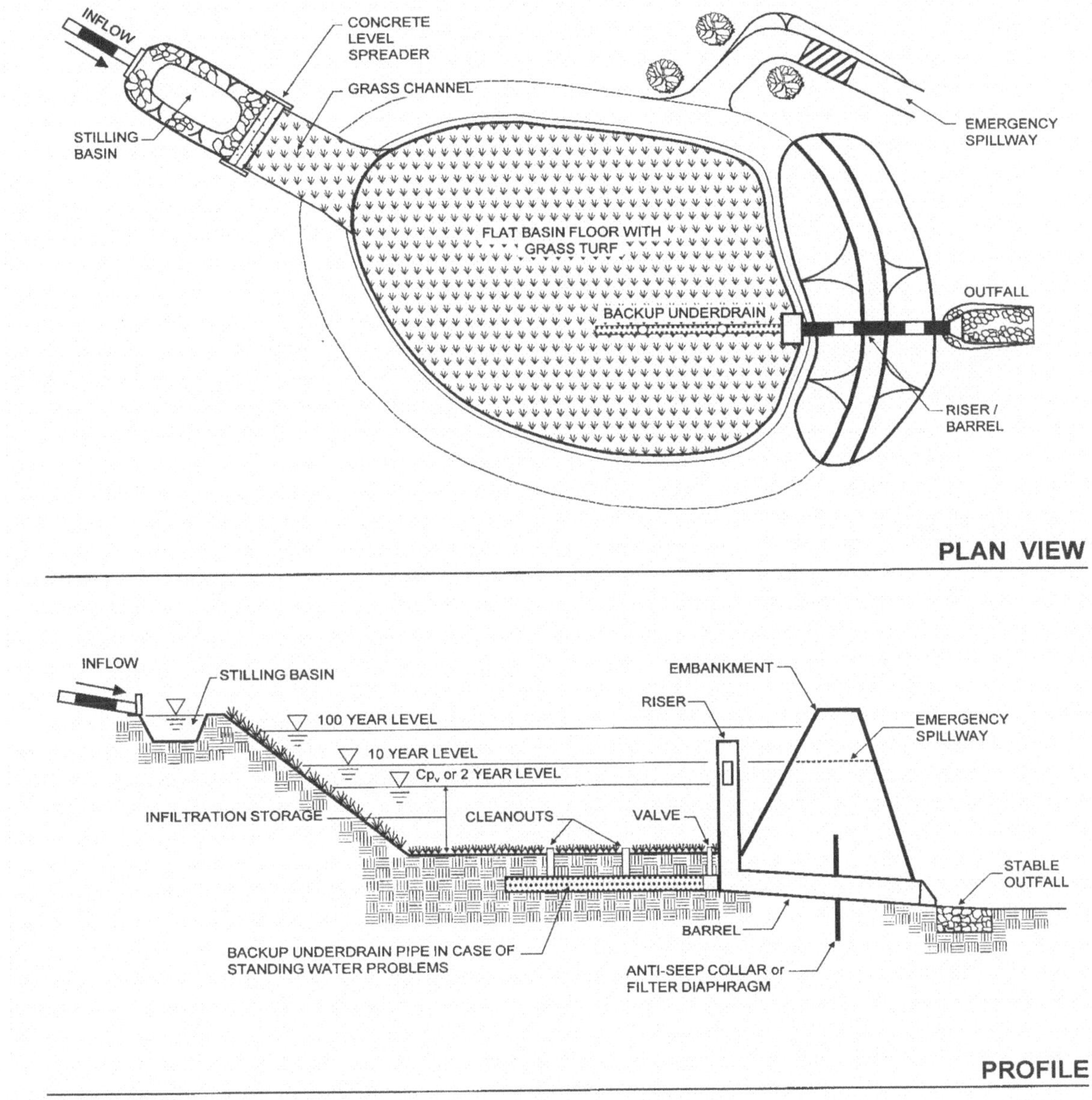

Figure 4-1. Schematic of Infiltration Basin.

4.3 Limitations

- Infiltration basins have potentially high failure rates, the main reasons being improper siting, design, and lack of maintenance. Proper pretreatment must be incorporated in the design and is essential for the longevity of this device.
- Soil conditions are the most critical factor for the site selection. In the U.S., a minimum infiltration rate of 0.5 in./hr is required and sites with Soil Types C and D (soil types with high clay content) are excluded. A minimum distance to the

groundwater table is required to ensure sufficient hydraulic gradient to meet the specified drain time.

- Infiltration basins are not appropriate for treating flows with significant loads of sediment and other pollutants due to the potential for clogging. They are also not appropriate for industrial sites or locations where spills may occur.
- Frequent inspections and maintenance have to be undertaken and an aggressive maintenance schedule must be adopted.
- Infiltration basins are not suitable on fill sites or steep slopes, or where there is a risk of groundwater contamination due to the soil structure. They should not be employed adjacent to drinking water wells for this reason.
- The proximity of infiltration basins to other structures, including foundation, must be taken into account during the design phase due to the risks of changes in soil structure as a result of the infiltration processes. Current U.K. Building Regulations advise that infiltration systems should not be constructed within 5 m (16 ft) of roads or building foundations.

4.4 Siting Criteria

The key element in siting infiltration basins is identifying sites with appropriate soil and hydrogeologic properties, which is critical for long term performance. In one study conducted in Prince George's County, Maryland (Galli, 1992), all of the infiltration basins investigated clogged within 2 years. This trend may not be the same in soils with high infiltration rates, however. A study of 23 infiltration basins in the Pacific Northwest showed better long-term performance in an area with highly permeable soils (Hilding, 1996). In this study, few of the infiltration basins had failed after 10 years.

The following guidelines for identifying appropriate soil and subsurface conditions should be followed:

- Determine soil type from mapping, soil survey tables, and onsite investigation to estimate permeability.
- Eliminate sites with high groundwater levels.
- Locate away from buildings, slopes, and highway pavement, as well as wells and bridge structures.
- Do not consider sites constructed of fill, having a base flow, or with steep slope.
- Ensure that adequate head is available to operate flow-splitter structures designed to divert the water quality volume to the infiltration basin.

Many guidelines recommend against the use of infiltration basins in catchments prone to high pollutant loads or spills of toxic materials because of potential groundwater contamination. However, EPA (1983) and Nightingale (1975; 1987a,b,c; 1989) found no evidence of transport of pollutants for any distance in the subsurface. For instance, a report by Pitt, Clark, and Parmer (1994) highlighted the potential for groundwater contamination from intentional and unintentional stormwater infiltration. That report recommends that infiltration facilities not be sited in areas where high concentrations are present or where there is a potential for spills of toxic material. Conversely, Schroeder (1995) reported that there was no evidence of groundwater impacts from an infiltration basin serving a large industrial catchment in Fresno, California.

4.5 Design Guidelines

Table 4-1. Design Guidelines.

Design Criteria	Current Guidance/Recommendations
Soil Characteristics	♦ Infiltration basins require an extensive geotechnical exploration to determine the subsurface profile and the hydraulic conductivity of the *in situ* soils. ♦ In the U.S. preconstruction soil infiltration rate in the soil should be between 13 mm/hr (0.51 in./hr) and 100 mm/hr (4 in./hr).
Groundwater Levels	♦ Historical well records and geotechnical investigations must also be evaluated to establish potential ground water levels. ♦ The invert of infiltration basins should be located at least 1.2 m (4 ft) above the seasonally high groundwater elevation.
Inlet Infrastructure	♦ Inflow to the basin may be by sheet flow down the sides, riprap/concrete chute for concentrated surface flows from paved areas, or a drain discharging from the basin side with suitable scour protection and headworks (CIRIA, 2000a,b,). ♦ Include energy dissipation in the inlet design for the basins to minimize scour potential. ♦ Avoid designs that include a permanent pool to reduce opportunity for standing water and associated vector problems.
Pretreatment	♦ Provide pretreatment if sediment loading is a maintenance concern for the basin. ♦ Inlet basin should be sized to hold the first flush of polluted flow.
Flood Control/ Treatment Volumes	♦ The required capture volume is determined by local regulations, i.e., in CA capture 85% of the event. ♦ Infiltration basins shall be designed to capture, store and treat (infiltrate) the water quality volume. ♦ Basin-sized so that the entire water quality volume is infiltrated within 48 hours. ♦ Infiltration/retention times longer than 48 hours should not be considered to minimize the potential for mosquito breeding. ♦ After a storm with a short recurrence interval (< 2 years), the infiltration of water should be sufficient to drain half the stored runoff within 24 hours. In extreme storms it will take longer to empty an infiltration basin (CIRIA, 2000a,b).
Basin Invert Area and Shape	♦ Basin invert area should be determined by the equation: ♦ $A = \frac{WQV}{kt}$ Where: A = Basin invert area (m^2) WQV = water quality volume (m^3) k = 0.5 times the lowest field-measured hydraulic conductivity (m/hr) t = drawdown time (48 hr) ♦ Infiltration basins should have gently sloping sides and a flat base area. ♦ Side slope should be 1 in 4 or shallower (this is considered to be the steepest slope which can easily be negotiated by someone trying to exit the basin and that can be mowed with a ride-on mower) (CIRIA, 2000a,b). ♦ There should be a smooth transition between the base and sides of the infiltration basin. ♦ Curving, irregular plan forms are preferred from an amenity and visual standpoint. ♦ Depths of construction generally range from 0.5–3.0 m (0.15–0.9 ft). A freeboard should be maintained above the maximum water level, but this should be based on the calculated risks of overtopping.
Vertical Piping	♦ In the U.S. the use of vertical piping, either for distribution or infiltration enhancement shall not be allowed to avoid device classification as a Class V injection well per *40 CFR* 146.5(e) (4).

Catchment Topography	♦ Land Slope: Infiltration basins can be located on slopes of up to 15%.
Outlet Infrastructure	♦ Infiltration basins shall have an overflow outlet or bypass to limit the risk of overtopping the device. Alternatively, detention storage can be built on top of the infiltration basin.
Vegetation	♦ The infiltration basin sides and bottom should be stabilized with vegetation or nonvegetative measures to minimize erosion and controls dust. Using vegetation at bottom can improve the infiltration characteristics.
Maintenance Needs	♦ Dedicated access to the basin bottom should be provided for maintenance vehicles. ♦ Basins may be lined with a layer of filter material such as coarse sand to prevent the buildup of impervious deposits on the natural soil surface.
Safety Measures	♦ Barrier planting (or a vehicle barrier if next to a road) should be considered where any significant pool of water is likely to remain following a storm. ♦ At infiltration basins used for a secondary purpose, notices should be erected at the site. In conjunction with education of the local residents, such measures should emphasize that the basin is primarily a drainage device and inform people how to use it safely.

4.6 Construction Guidelines

- Before construction begins, stabilize the entire area draining to the facility. If impossible, place a diversion berm around the perimeter of the infiltration site to prevent sediment entrance during construction. Stabilize the entire contributing drainage area, including the side slopes, before allowing any runoff to enter once construction is complete. The vegetation must be allowed to become fully established before the basin comes into use.
- Place excavated material such that it cannot be washed back into the basin if a storm occurs during construction of the facility.
- Build the basin without driving heavy equipment over the infiltration surface. Any equipment driven on the surface should have extra-wide ("low pressure") treads or tires. Prior to any construction, rope off the infiltration area to stop entrance by unwanted equipment.
- After final grading, till the infiltration surface deeply.
- Use appropriate erosion control seed mix for the specific project and location.
- Topsoil should be uniformly distributed and lightly compacted to a depth of 100 mm (4 in.) (U.K. guidance). Topsoil should not be placed while frozen or waterlogged, or when the sub-grade is excessively wet. The application of seed and fertilizer should be as uniform as possible (CIRIA, 2000a,b).
- To prevent the topsoil becoming dry during establishment of the vegetation, it might be necessary to irrigate the infiltration basin.

4.7 Performance

As water migrates through porous soil and rock, pollutant attenuation mechanisms include precipitation, sorption, physical filtration, and bacterial degradation. If functioning properly, this approach is presumed to have high removal efficiencies for particulate pollutants and moderate removal of soluble pollutants. Actual pollutant removal in the subsurface would be expected to vary depending upon site-specific soil types.

This technology eliminates discharge to surface waters except for the very largest storms; consequently, complete removal of all stormwater constituents can be assumed.

Performance (relative to surface water discharge):

Sediment:	100%	Litter:	100%
Total Nitrogen:	100%	Oxygen Demanding Substances:	100%
Total Phosphorus:	100%	Oil & Grease:	100%
Dissolved Zinc:	100%	Bacteria:	100%
Toxic Materials:	100%		

Qualitative performance information has been gleaned from some parts of the world, as indicated below:

4.7.1 California

Infiltration basins have a long history of use in California, especially in the Central Valley. Basins located in Fresno were among those initially evaluated in the National Urban Runoff Program, and were found to be effective at reducing the volume of runoff, while posing little threat to groundwater quality (U.S. EPA, 1983). Proper siting of these devices is crucial as underscored by the experience of Caltrans in siting two basins in Southern California. The basin with marginal separation from groundwater failed immediately and could never be rehabilitated.

4.7.2 Bordeaux, France

An infiltration basin, holding 80,000 m^3 over an area of some 2 ha (65 acre-feet), was constructed at Venissieux in 1975 (Chocat, Barraud, and Alfakih, 1999). Inflow to the basin first passed through a sand trap before entering the basin, in which the base was covered with a geotextile and 400 mm (1.3 ft) sand. In 1988 the basin was divided into two sections and in 1990 the geotextile and sand layer were renewed. The basin now comprises sand traps, a settling basin of 1.1 ha (2.7 ac), from which flow passes through a flow regulator and hydrocarbon separator before entering an infiltration basin of 0.9 ha (2.2 ac). The Venissieux catchment is 380 ha (940 ac) of mixed industrial, agricultural and residential land use. Groundwater is 3–5 m (9.8–16 ft) below the base of the basin. Water samples have been collected at the inlets to the basin; at the outlet from the settling basin and in the infiltration basin, before and after the hydrocarbon separator.

In addition, soil samples were collected and cores obtained through the base of the infiltration basin to a depth of 1.5 m (4.9 ft). The settling basin has achieved decreases in suspended solids of between 20% and 70%. Examination of the soil cores showed that solids less than 2 mm (0.79 in.) particle size were seriously contaminated by heavy metals and mineral oils, although this decreased with depth until between the base of the geotextile and 0.5 m (1.6ft) into the subgrade the pollution was slight, despite the 20 years of infiltration. No data was reported on the possible impact on groundwater quality.

4.7.3 United Kingdom

Two systems were identified in the United Kingdom by Pratt (1995), but they have not been monitored. The infiltration basin at Ipswich has been in existence for some 25 years with little or no maintenance being conducted. The Yellowham Hill basins, which were constructed in 1991 as part of a road improvement scheme, are situated over an aquifer and could provide valuable data on the possible impact of highway runoff (Codling, 1992). Unlike many of the other forms of SUDS implemented in Scotland in recent years, retention basins and other infiltration devices have not been employed because ground conditions are generally unsuitable. In England there are extensive regions of the country that would be appropriate for this type of drainage.

4.8 Maintenance

Table 4-2 outlines the main maintenance requirements of an infiltration basin.

Table 4-2. Typical Maintenance Activities for Infiltration Basins (after WMI, 1997).

Activity	Schedule
♦ Inspect facility for signs of wetness or damage to structures ♦ Note eroded areas. ♦ If dead or dying grass on the bottom is observed, check to ensure that water percolates 2–3 days following storms. ♦ Note signs of petroleum hydrocarbon contamination and handle properly.	Semi-annual inspection
♦ Mow and remove litter and debris. ♦ Stabilize eroded banks. ♦ Repair undercut and eroded areas at inflow and outflow structures.	Standard maintenance (as needed)
♦ Disc or otherwise aerate bottom. ♦ Dethatch basin bottom.	Annual maintenance
♦ Scrape bottom and remove sediment. Restore original cross-section and infiltration rate. ♦ Seed or sod to restore ground cover.	Five-year maintenance
♦ Remove sediment from the forebay.	When the accumulated sediment volume exceeds 50% of forebay capacity.

4.9 Design Life

The design life of infiltration basins is highly variable depending on soil and subsurface conditions.

4.10 Costs

Infiltration basins are relatively cost-effective practices because little infrastructure is needed when constructing them. One study estimated the total construction cost at about \$2/ft^3 (\$71/m^3) (adjusted for inflation) of storage for a 0.25-ac (0.01-ha) basin (SWRPC, 1991). As with other BMPs, these published cost estimates may deviate greatly from what might be incurred at a specific site. For instance, Caltrans spent about $18/ft^3 (640/m^3) for the two infiltration basins constructed in southern California, each of which had a water quality volume of about 0.34 acre-feet (420 m^3).

Infiltration basins typically consume about 2%–3% of the site draining to them, which is relatively small. Maintenance costs are estimated at 5%–10% of construction costs.

One cost concern associated with infiltration practices is the maintenance burden and longevity. If improperly maintained, infiltration basins have a high failure rate. Thus, it may be necessary to replace the basin with a different technology after a relatively short period of time.

Maintenance costs of infiltration basins consist almost exclusively of vegetation management, unless clogging occurs.

4.11 Research Needs

Significant uncertainty exists as to the fundamental feasibility of the infiltration approach. The basic issue is whether the temporary but deep ponding in basins makes clogging inevitable. Other uncertainties include:

- Pretreatment system that can ensure basin longevity;
- The impact of basins in freezing conditions and during snowmelt conditions;
- The accuracy of current field methods for estimating long-term local soil infiltration rates; and
- The actual monitored pollutant removal performance of a fully operational basin, particularly with respect to soluble nutrients.

4.12 Additional Resources

Center for Watershed Protection (CWP), 1997, *Stormwater BMP Design Supplement for Cold Climates.* Prepared for U.S. Environmental Protection Agency Office of Wetlands, Oceans and Watersheds. Washington, DC.

Ferguson, B.K., 1994, *Stormwater Infiltration.* CRC Press, Ann Arbor, MI.

Stormwater Management Resource Centre (SMRC), 2002, [http://www.stormwatercenter.net/] Accessed August 19, 2002.

U.S. Environmental Protection Agency (U.S. EPA), 1983, *Results of the Nationwide Urban Runoff Program: Volume 1—Final Report*, Report WH-554, Water Planning Division, Washington, D.C.

U.S. Environmental Protection Agency (U.S. EPA), 1993, *Guidance to Specify Management Measures for Sources of Nonpoint Pollution in Coastal Waters.* EPA-840-B-92-002. U.S. Environmental Protection Agency, Office of Water, Washington, D.C.

CHAPTER 5.0

INFILTRATION TRENCHES/SOAKAWAYS/ FILTER DRAINS

5.1 Description

Infiltration trenches are shallow excavations that are filled with rubble or stone to create temporary subsurface storage for infiltrations of stormwater runoff. Some manufacturers also have developed modular plastic systems which provide greater void space than rock filled trenches and reduce the amount of excavation required to store the water quality volume. Usually a small portion of the runoff, the first flush, is diverted to the infiltration trench. The runoff gradually filters through the bottom and/or sides of the trench into the subsoil and eventually to the water table. The primary pollutant removal mechanism is via filtering through the soil. Figure 5-1 provides a schematic of a typical end-of pipe system. Trench designs may be modified to include vegetative cover and other features, establishing a biofiltration area.

Infiltration trenches may be designed for complete exfiltration or partial exfiltration where the first flush of the runoff volume is routed to the trench and the remainder is bypassed and conveyed to additional BMPs.

Infiltration trenches should always be constructed with pretreatment. The use of infiltration technologies should be avoided in areas with high potential pollutant loading. In groundwater drinking supply recharge areas (Zone II and Interim Wellhead Protection Areas (IWPA)) infiltration technologies may be used for uncontaminated rooftop runoff only (Stormwater Management, 1997).

In the United Kingdom, infiltration trenches are commonly known as soakaways and these can take different forms and sizes. They are built as square or circular pits, which are filled with rubble or lined with dry-joined brickwork or precast, perforated concrete ring units, surrounded by granular backfill. They can also take the form of trenches that follow convenient contours (BRE, 1991). Extensive use has been made of soakaways in draining small catchments in remote and rural areas.

Filtration trenches or filter trenches (Figure 5-2), which include underdrain systems in their design, combine some of the attributes of infiltration trenches and filter systems. They are used in the United Kingdom in areas of low permeability soils with the main function of purifying water prior to conveying it downstream. Filter trenches with impermeable liners are also used in areas where the risk of groundwater contamination is high.

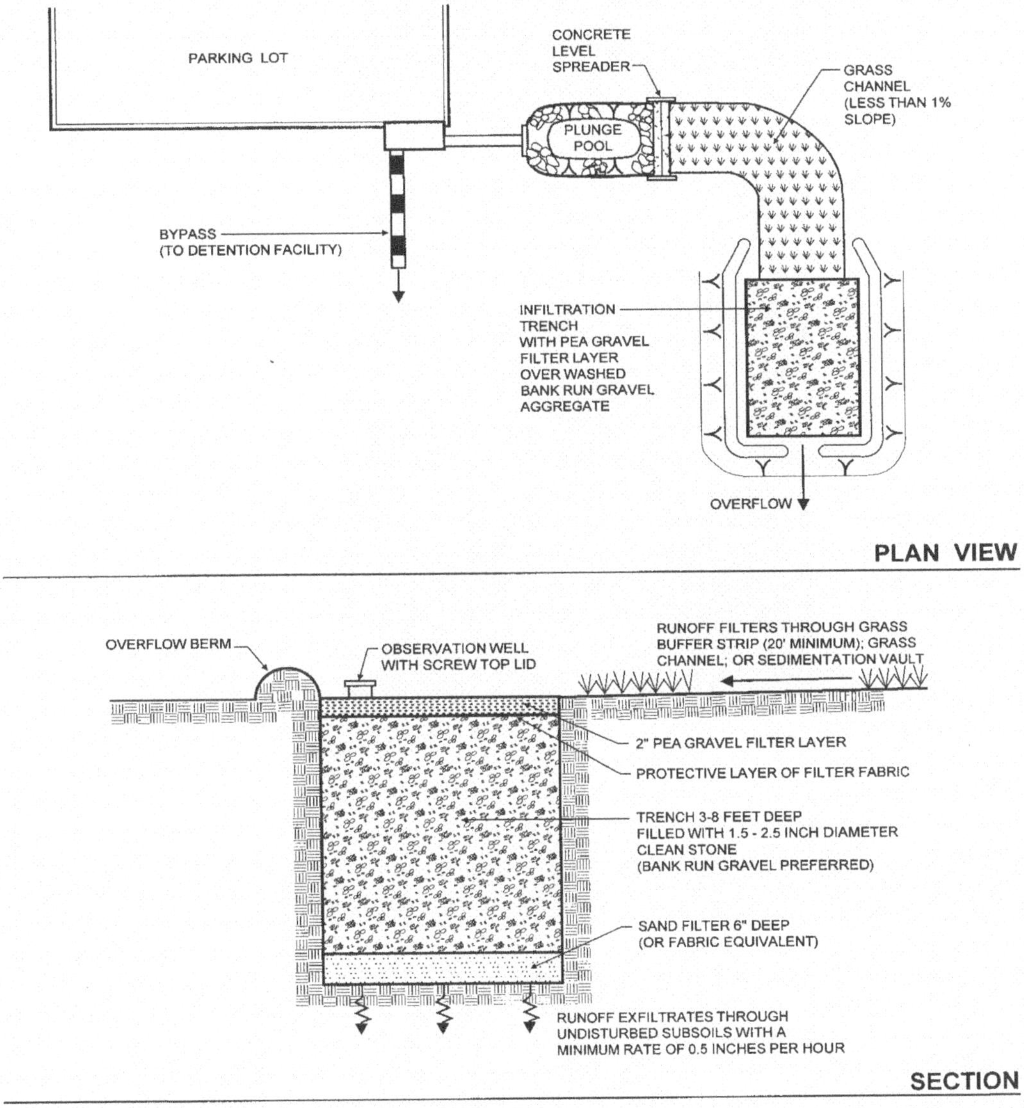

Figure 5-1. Schematic of Infiltration Trench as End-of-Pipe System (Maryland Department of the Environment, 2000).

Figure 5-2. Filter Drain.

Perforated pipes buried within the gravel of a filter trench are often used to distribute the inflow runoff along the length of the trench (Duchene and McBean, 1992).

Figure 5-3 shows a typical arrangement of this application.

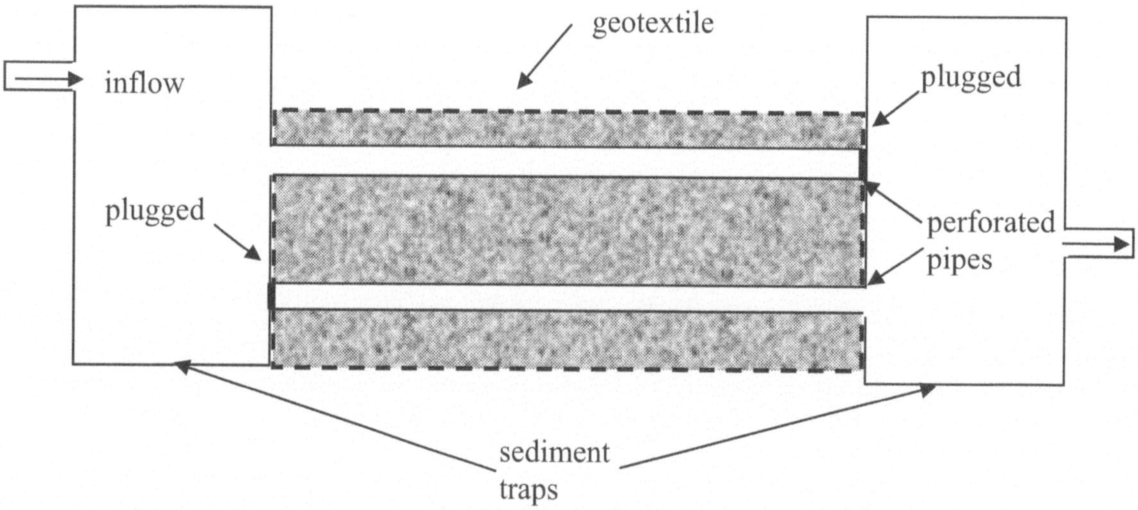

Figure 5-3. Schematic of Filter Trench with Perforated Pipes as End-of-Pipe System.

In the application of roadside drainage, a perforated pipe in combination with an infiltration/filtration trench is known as filter drain. The usual arrangement is in combination with a trapped gullypot.

Figure 5-4 shows a schematic of a typical road drainage system. It is estimated that filter drains are used to collect surface water runoff from 25% of all major roads in the United Kingdom (CIRIA, 1999). A similar system, termed a partial exfiltration trench, was described in the United States by Sansalone, Smithson, and Koran (1998).

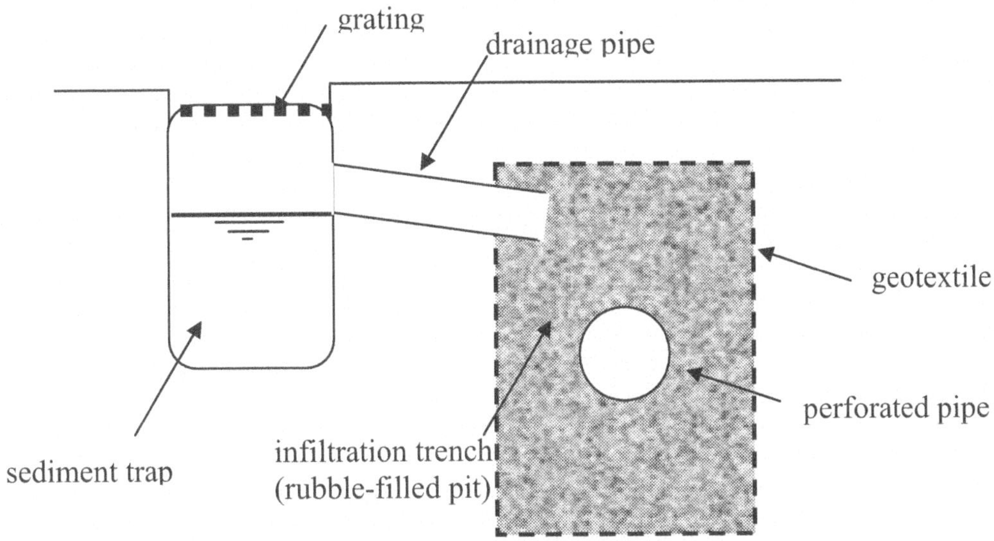

Figure 5-4. Schematic of a Filter Drain for Road Drainage (modified from Price, 1994).

5.2 Advantages

- Such systems can provide 100% reduction in the load discharged to surface waters.
- An important benefit of infiltration trenches is the approximation of predevelopment hydrology during which a significant portion of the average annual rainfall runoff is infiltrated and evaporated rather than flushed directly to creeks.
- If the water quality volume is adequately sized, infiltration trenches can be useful for providing control of channel forming (erosion) and high frequency (generally less than the 2-year) flood events.
- As an underground BMP, trenches are unobtrusive and have little adverse impact on site aesthetics.

5.3 Limitations

Although infiltration trenches/filter drains can be a useful management practice, they have several limitations, including the following (U.S. EPA, 1997):

- High failure rate due to poor maintenance, improper siting, or high debris input;
- May not be appropriate for industrial sites or locations where spills may occur;
- Infiltration trenches are limited to relatively small catchments;
- Infiltration trenches and soakaways require a minimum soil infiltration rate or maximum emptying time depending on the location (not the case for filter drains and filter trenches);
- Cannot be used in areas of high groundwater level or close to building foundations;
- Not suitable for fill sites or steep slopes;
- Risk of groundwater contamination;
- Upstream drainage area must be completely stabilized before construction;
- Cost of replacing filter material once blocked; and
- No visual enhancement.

5.4 Siting Criteria

The use of infiltration trenches may be limited by a number of factors including type of native soils, climate, and high groundwater tables. Site characteristics, such as excessive slope of the drainage area, fine-grained soil types, and near-surface location of the water table and bedrock, may restrict the use of infiltration trenches. Generally, infiltration trenches are not suitable for areas with relatively impermeable soils containing clay and silt or in areas with fill.

However, infiltration trenches were successfully implemented in Sweden in soil of a boulder clay of impermeable nature. The benefit from the use of the devices was in maintaining an adequate level of soil moisture, thus preventing consolidation of the soil that could result in unacceptable settlement of buildings. Holmstrand (1984) reported that up to two-thirds of the water discharged to the trenches was not passed to the outfall and that losses were due to evaporation from the soil and plant transpiration.

As stated earlier, the potential for groundwater contamination must be carefully considered, especially if the groundwater is used for human consumption or agricultural purposes. Serious concern has risen with regard to groundwater contamination from highway runoff draining into filter trenches (Price, 1994). In Germany highway runoff is generally not permitted to be directly collected in infiltration trenches due to the risk of groundwater contamination (ATV 138, 2001).

The infiltration trench is not suitable for sites where hazardous spills are likely to occur, such as industrial sites. In these areas, other BMPs that do not allow interaction with the groundwater should be considered.

Soil infiltration rate should be at least 13 mm/hr (0.5 in./hr),and the water table depth should be at least 3 m (9.8ft) below the trench invert for complete infiltration of the stormwater volume (Caltrans, 2002).

5.5 Design Guidelines

Observations from Tokyo, Japan (Haneda, Kagawa, and Maruyama, 1996) suggest that such infiltration trench/filter drain systems may operate satisfactorily with significant benefits in the reduction of direct discharges to watercourses. A key factor in continued satisfactory operation is the design, operation, and maintenance of inlet structures to the underground devices. Where roof waters discharge into sediment/debris traps, the underground, stone-filled trenches retained their infiltration capacity after 11 years of service. However, where the waters entered the system from paved surfaces, carrying considerably more silt and debris, the infiltration rates fell rapidly due to blockage (Pratt, 2001).

This confirmed the previous findings by Minagawa (1990). Excavations of infiltration trenches revealed less silt within the stone fill and on the base of the trenches than expected. Extensive use was made onsite of U-shaped collecting channels and of silt traps: these played an important role in the pretreatment of materials before entry to the infiltration trenches.

U.K. design guidelines are outlined in the BRE 365 (1991) and the CIRIA reports C521 and C522 (2000a,b). German guidelines are outlined in the ATV 138 (2001), and various guidelines have been published in the United States (Schueler, 1987; Stormwater Management Volume Two, 1997; U.S. EPA, 1999c; MDE, 2000). Specific designs may vary considerably, depending on site constraints or preferences of the designer or community. There are some features, however, which should be incorporated into most infiltration trench designs, and these are discussed in Table 5-1.

Table 5-1. Design Guidelines.

Design Criteria	Current Guidance/Recommendations
Hydraulic Residence Time	♦ Typically, filter drains are designed to store 10 mm (0.03') of rainfall from the contributing area. Storage is provided below the high level outlet pipe. The low-level (under-) drain is sized on the basis that it will discharge the treatment volume in 24 hours. If the drain is near a road and the subgrade is susceptible to waterlogging, the drainage time should be reduced.
Granular Fill	♦ Specify locally available granular fill material, e.g., trench rock that is 40–60 mm (1.5–2.5in.) in diameter. ♦ A perforated or porous distributor pipe can be included in the design. ♦ For filter drains, this pipe should only be provided over the last few meters before the outlet, or adjacent to a manhole. The absence of a continuous drain for flows up to the design volume increases the attenuation and the potential biological treatment (CIRIA, 2000a,b). ♦ With long infiltration trenches it is advisable to provide inspection tubes at regular intervals along the trench. The extreme ends of the trench should be identified by inspection tube covers or other access covers. These provide confirmation of the line and may be convenient as points of inflow for intermediate connections. (CIRIA, 2000a,b). ♦ A high level overflow pipe should be provided for filter drains and considered for infiltration trenches. An alternative method might be to allow the drain to back up and overflow, but the route of the floodwater should be carefully designed to avoid adverse impacts, especially frequent waterlogging of the road subgrade.
Inspection Tubes	♦ An inspection tube (e.g., 225 mm [0.74 ft] perforated pipe) should be directly connected to any incoming drains to allow the free overflow of water and prevent any build-up of debris at the end of the pipe. Debris may be removed periodically from the base of the inspection tube. The tube should be slotted or perforated to minimize the build-up of hydraulic head (CIRIA, 2000a,b,).
Pretreatment Infrastructure	♦ Pretreatment structures, such as a vegetated buffer strip or water quality inlet, can increase longevity by removing sediments, hydrocarbons, and other materials that may clog the trench. In the United Kingdom, typically trapped gully pots are used in the application of road drainage and extensive use has been made of catch basins. ♦ Infiltration trenches should always be constructed with pretreatment to reduce the sediment load. BRE 365 (1991) suggests the use of wet sumps and T-piece inlets to distributor pipes in trench systems.
Subsoil	♦ The soil permeability is often difficult to assess. Some countries use data tables for each soil type, other countries require onsite investigations and the Dutch procedure allows both. ♦ A hydrogeological and a geotechnical evaluation should be undertaken to determine the suitability of the site for infiltration drainage. This is particularly important on sites where there is filled ground, as the frequent discharge of additional waters might change the soil characteristics, either chemically or structurally (CIRIA, 2000a,b). ♦ The soil infiltration rate should be determined at the location and depth of the proposed device (CIRIA, 2000a,b).
Water Table	♦ The site should be checked with respect to groundwater contamination vulnerability, to ensure that infiltration is an acceptable method of disposal. ♦ Infiltration systems must not adversely affect groundwater resources. ♦ Infiltration devices should not be constructed in ground where the water table reaches close to the bottom of the device at any time of the year.
Observation Well	♦ Provide observation well to allow observation of drain time.

Design Criteria	Current Guidance/Recommendations
Trench Lining	♦ The sides and the bottom of the trench should be lined with permeable, geotextile fabric. A layer of fine sand (clean, fine aggregate) may be substituted or used in addition on the bottom. ♦ Geotextiles should be selected to suit the surrounding soil particle size and permeability (CIRIA, 2000a,b).
Infiltration Area	♦ Determine the effective area. In the United States this is defined as the bottom surface area needed to drain the trench within 72 hours by dividing the treatment volume by the infiltration rate. In the United Kingdom this is calculated using 50% of the total depth, excluding the bottom area. The trench should be half-drained within 24 hours. ♦ Another parameter of great diversity is the effective percolation area. In many countries the bottom of the facility is assumed to be clogged after some time and is therefore not taken into consideration. Most countries use 50% of the side walls. Although this a time dependent variable it is used as a constant for the design approach. None of the procedures acknowledge the change of head (Monster and Leeflang, 1996).
Trench Volume	♦ Determine the trench volume. This depends on whether the trench is designed to treat the first flush or the total runoff volume. The defined volume is assumed to fill the void space based on the computed porosity of the rock or other matrix.

5.6 Construction Guidelines

- Soakaways and filter trenches should not be used for untreated drainage of construction sites, where runoff is likely to contain large amounts of silt, debris, and other pollutants.
- Inspection tubes and/or covers should be convenient for access and the provision of an easement should be considered where multiple properties discharge to a single soakaway.
- Depending on soil type and site location, the trench may be lined with a geotextile. This should wrap over the top of the infill material to prevent ingress of soil or other surfacing material. Wherever the trench extends to the ground surface, the geotextile should be protected by 150 mm of granular or filler material.

5.7 Performance

Despite the extensive use that has been made of infiltration pits and trenches or soakaways there has been only limited examination of their performance (Warnaars et al., 1999). There has been widespread concern about the hydraulic performance of these devices, with general expectation that failure through blockage by silt and debris, would necessitate reconstruction within a limited time period (Pratt, 2001).

Infiltration trenches function similarly to rapid infiltration systems that are used in wastewater treatment. Estimated pollutant removal efficiencies from wastewater treatment performance and modeling studies shown in Table 5-2 can be used to estimate the removal of pollutants entering the subsurface. Based on these data, infiltration trenches can be expected to remove up to 90% of sediment, metals, coliform bacteria and organic matter, and up to 60% of phosphorus and nitrogen in the runoff (Schueler, 1992). Biochemical oxygen demand (BOD) removal is estimated to be between 70% and 80%. Lower removal rates for nitrate, chlorides, and soluble metals should be expected.

Table 5-2. Typical Removal Rates (Schueler, 1992).

Typical Percent Removal Rates	
Sediment	90
Total Phosphorus	60
Total Nitrogen	60
Metals	90
Bacteria	90
Organics	90
Biochemical Oxygen Demand	75

There is considerable variability in the effectiveness of infiltration trenches, and proper siting, design and maintenance improves their performance.

In the United States in the late 1990s, research was undertaken on the water quality improvement capability of a partial infiltration trench at a site on the Millcreek Expressway in Cincinnati, Ohio (Sansalone, Smithson, and Koran, 1998). Road runoff was intercepted by a small section of porous pavement and filtered through the granular backfilled trench. Runoff could infiltrate the adjacent soil or be discharged from the filter media via a pipe, the flow from which was monitored and sampled. The heavy metal and total suspended solids removal efficiency results showed that the system was an effective trap, as shown in Table 5-3. Concern was expressed about the potential for clogging of the system and the effect this would have on the design life of 10 years.

Table 5-3. Typical Removal Rates (Sansalone, Smithson, and Koran, 1998).

Typical Percent Removal Rates	
Dissolved Metals:	
- Zinc	>95%
- Copper	>85%
- Cadmium	>80%
- Lead	70%-95%
Particulate-bound metals:	
- Zinc	75%-95%
- Copper	85%-95%
- Cadmium	79%-90%
- Lead	85%-95%

Minagawa (1990) reported field observations of the silting of sedimentation/infiltration chambers and of porous pipes within interconnecting infiltration trenches over 5 years at three residential areas in Tokyo. The infiltration capacity of the chambers was effectively reinstated after removal of 150 mm (0.5 ft) of the crushed rock and deposits at the bottom of the chambers (Pratt, 2001).

A study of the impact of highway runoff on soil and groundwater conducted in Switzerland provides information on likely infiltration water quality below infiltration surfaces (Mikkelsen et al., 1997). Two sites were studied. At the first, highway runoff was discharged via a pipe from a curb inlet gully into a depression on the grass-covered verge (shoulder). The road carried an estimated 37,000 vehicles a day in 1993, and the drainage had been in service since 1959. High concentrations of heavy metals, a number of polynuclear aromatic hydrocarbons (PAHs) and adsorbed organically bound halogens were found in the upper 500 mm (1.6 ft) of runoff sludge and soil, but the concentrations decreased rapidly to background levels. At the second site, highway runoff was discharged into 3 m deep soakaways at a location where traffic density was estimated at 2300 vehicles per day. Here, the runoff sludge was some 200 mm (0.66 ft) deep, and no traces of contamination were found further than another 200 mm (0.66 ft) below that level. Similar findings were reported for two soakaways investigated at Brandon, Suffolk (Pratt, 1995, 1996). High concentrations of total organic carbon and heavy metals were associated with fine, organic material accumulation in the first 400 mm (1.3 ft) of sediment in the base of the soakaways. Below 400 mm the pollutant levels appeared to approach the background levels.

It is therefore proposed that the formation and continued presence of a layer of sludge at the base of the soakaway is important in retaining pollutants through filtration and adsorption, resulting in significant build-up of copper, zone, cadmium, PAHs, and halogens in that layer (Pratt, 2001). It was concluded at the Swiss sites that "leaching of heavy metals is limited and that contamination of potable groundwater with metals is of little practical concern within a reasonable time frame." This conclusion holds for the PAHs, which are known to adsorb well in soil systems, and for the type of pollutants included in the organically bound halogen analysis (Pratt, 2001). However, it is stressed by Mikkelsen that soluble components such as pesticides and deicing salts may pass directly through infiltration systems.

A recent study at HR Wallingford looked at the hydraulic performance of a perforated concrete ring soakaway (Abbott and Comino-Mateos, 2001). Observations of actual infiltration from the soakaway, occurring within 24-hour periods during different storm events, showed values of 6.4% to 8.9% of fill volume instead of the 50% recommended by BRE and CIRIA.

U.K. research on a typical roadside filter drain was conducted in 1999 (Schlüter, Spitzer, and Jefferies, 2002). Preliminary results after 2 years of operation showed peak TSS removal rates of 75% but conductivity increasing by 73%. The conductivity rise was thought to be due to the use of roadside salts. Percentage outflow rates were found to be in the order of 40%, and lag times varied from 0 to 11.5 hours. Analysis of the most recent datasets suggests sediment break-through at the outflow of the system. Monitoring is ongoing. Water quality performance measurements are summarized in Table 5-4.

Table 5-4. Trench Pollutant Removal Efficiency Data.

Study	Removal Efficiencies (% Removal)						Type
	TSS	o-phos	TN	NO_3	Metals	HCs	
Folkes-Skinner, 2002 (Hopwood Services)	85			-86	Cu: 72 Pb: 85 Zn: 83 Cd: 38		Filter drain
Perry and McIntyre, 1986	85				Pb: 83 Zn: 81	70	Filter drain
Winer, 2000		100		42			Infiltration trench
Schueler, 1992	90	60	60		90		Infiltration trench
Sansalone, 1999					Cu: 85 Pb: 85 Zn: 85 Cd: 80		Infiltration trench
Schlüter, Spitzer, and Jefferies, 2002	75						Filter drain

5.8 Maintenance

Longevity can be increased by careful geotechnical evaluation prior to construction and by designing and implementing an inspection and maintenance plan.

Price (1994) points out that gully pots discharging to filter drains were formerly emptied every two months. Since the reorganization of the U.K. local government in 1974, the interval between gully emptying has increased to a year or more. Data published by CIRIA (1995) show gullypot filling times of approximately 4 to 6 months. Clearly at these locations an annual emptying interval would not be sufficient. There is concern that water held in traps for a long time may provide a breeding ground for bacteria and concentrate pollutants which are then carried out as a concentrated first flush when it rains (CIRIA, 1999; Price, 1994, Butler and Memon, 1999).

A problem frequently encountered with drains and sewers is the ingress of tree roots through poor cracks in the pipework. Roots will be drawn to the location of a soakaway. It is unclear whether this is a benefit or a problem as the roots may provide additional pathways for water to enter the soil, and the tree will take up water itself.

The following points should be considered (U.S. EPA, 1999c):

- Inspections and maintenance for standing water, trash and debris, and sediment accumulation.
- Observe drain time for the design storm after completion or modification of the facility to confirm that the desired drain time has been obtained (BRE, 1991).
- An aggressive maintenance and inspections schedule is advisable to ensure longevity of the filtration/infiltration device. The schedule should vary with location and system used.
- Remove accumulated trash and debris in the trench as often as practicable.
- If sediment is visible on top of the trench, remove top layer of trench, silt, filter fabric and stone; wash stone and reinstall fabric and stone into trench.
- Inspect for standing water after main rainfall events.

Typical maintenance activities are summarized in Table 5-5.

Table 5-5. Typical Maintenance Activities for Infiltration Trenches/Soakaways.

Activity	Schedule
♦ Removal of debris from the floor of the inspection tube or chamber ♦ Cleaning of any filters on downpipes ♦ Emptying of debris from any vortex filter trap or catchpit. ♦ Removal and washing of exposed stones on the trench surface. ♦ Trimming of any roots that may be causing blockages.	Annually (semi-annually the first year)
♦ Replacement of the geotextile and rock fill will be necessary if the device becomes blocked with silt (> 5 years).	As needed (infrequent)

5.9 Design Life

Infiltration practices have historically had a high rate of failure. One study (Galli, 1992) conducted in Prince Georges County, Maryland, revealed that fewer than half of the infiltration trenches investigated (of about 50) were still functioning properly, and less than one-third still functioned properly after 5 years. Many of these practices, however, did not incorporate advanced pretreatment. By carefully selecting the location and improving the design features of infiltration practices, their performance should improve (U.S. EPA, 1999c).

EPA (1999c) suggests design lives of between 5 and 15 years. Geotextiles are likely to require replacement after 5 years. FWYA (1999) suggests design lives of between 10 and 15 years.

5.10 Costs

5.10.1 Construction Costs

Construction costs include clearing, excavation, placement of the filter fabric and stone, installation of the monitoring well, and establishment of a vegetated buffer strip. Additional costs include planning, geotechnical evaluation, engineering, and permitting. Typical construction costs, including contingency and design costs, are about \$5 per ft^3 (\$180/m^3) of stormwater treated (SWRPC, 1991; Brown and Schueler, 1997).

The Southeastern Wisconsin Regional Planning Commission (SWRPC, 1991) has developed cost curves and tables for infiltration trenches based on 1989 dollars. The 1993 construction cost for a relatively large infiltration trench [1.8 m (6 ft) deep, 1.2 m (4 ft) wide, with a volume of 68 m^3 (2400 ft^3) ranged from \$8000 to \$19,000. A smaller infiltration trench (0.9 m [3 ft] deep and 1.2 m [4 ft] wide, with a volume of 34 m^3 [1200 ft^3]) was estimated to cost from \$3000 to \$8500.

EPA (1999c) provides Table 5-6 on construction costs.

Table 5-6. Infiltration Trench Construction Costs (US EPA, 1999c).

SUD	Construction Cost Equation	Construction Costs	Typical Design Contingency and Other Capital Costs (30% of construction costs)	Annual Maintenance Costs (12 % of construction costs)
Infiltration Trench (SWRP, 1991; Schueler, 1987; U.S. EPA, 1993)	3.9 × water quality volume + 2900	\$47, 100	\$14,100	\$5, 600

Trench-type soakaways are less expensive than the alternative forms of ring unit soakaway, since the volumes of excavation and granular material backfill alone are similar for both types without even taking into consideration the cost of ring units. (Pratt, 1996).

In Europe, Hauger et al. cite cost functions of 200 Euro/m^3 for an infiltration trench. In the United Kingdom there is very little information, although figures of £50/m for a lined trench and £40/m for an unlined trench have been suggested.

5.10.2 Maintenance Costs

Maintenance costs include buffer strip maintenance and trench inspection and rehabilitation. SWRPC (1991) has also developed maintenance costs for infiltration trenches. Based on the above examples, annual operation and maintenance costs would average \$700 for the large trench and \$325 for the small trench. Typically, annual maintenance costs are approximately 5% to 10% of the capital cost (Schueler, 1987). Trench rehabilitation may be required every 5 to 15 years. Cost for rehabilitation varies according to site conditions and the degree of clogging. Trench rehabilitation is estimated to cost between 15% and 20% of the original capital costs (SWRPC, 1991). In the United Kingdom, there is little or no information on maintenance costs, although £30/m for lined trenches and £20/m for unlined trenches have been suggested. These costs do not include stone replacement and cleaning every 10 to 15 years.

5.11 Research Needs

5.11.1 Siting Criteria

There are substantial differences in siting criteria applied in Germany, the United Kingdom, and the United States regarding the potential for groundwater contamination from certain land uses, such as highways and industrial sites, despite research that indicates little transport of pollutants in the subsurface. Some additional research may be warranted to determine if there is a substantial threat from spills or runoff with relatively high pollutant concentrations.

5.11.2 Design Guidelines

Design methods vary significantly in how the required size of the trench is calculated based on permeability and surface available for infiltrations. Confirmation of modeling results with respect to effective drainage area (bottom and sides of trench) and effect of blockage of the bottom of the trench would be useful.

5.11.3 Performance

The potential impact on groundwater contamination remains a concern despite evidence that indicates rapid attenuation of pollutant concentrations in the subsurface. Clearly the amount of attenuation is strongly dependent on the characteristics of the soil, so additional evaluation of the performance in sandy soils near the water table may be warranted.

Little of the available performance data are from cold weather climates, so there is a large uncertainty in their performance in subfreezing weather and during snowmelt runoff conditions. Additionally, an investigation of possible groundwater contamination by soluble substances de-icing salts is required.

5.11.4 Construction Guidelines and Cost

The wide range of cost data indicates that much more information is needed in this area. Do the differences in cost reflect different regulatory requirements, differences in construction techniques, or perhaps site suitability? Identifying the primary elements that drive construction costs upward is a high priority and could help reduce the initial cost of this BMP/SUDS.

5.11.5 Maintenance Guidelines and Cost

Maintenance requirements for the trench itself are generally minimal, with the most significant maintenance needed for the pretreatment systems (e.g., buffer strip). A major question is whether the cost of reconstruction of the trench every 10 years will be substantially the same as the original construction cost.

5.12 Additional Resources

Schueler, T.R., and Lugbill, J., 1990, *A Current Assessment of Urban Management Practices: Techniques for Reducing Non-Point Source Pollutants in the Coastal Zone*. Department of Environmental Programs, Metropolitan Washington Council of Governments, Washington, D.C.

Todorovic, Z., Butler, D., Maksimovic, C., Pokrajac, D., and Deletic, A., 2002, Simulation of the long-term hydraulic behavior of stormwater infiltration trenches. Paper to be published in *Urban Water Journal*, Elsevier Science.

CHAPTER 6.0

RETENTION/IRRIGATION

6.1 Description

Retention/irrigation refers to the capture of stormwater runoff in a holding pond and subsequent use of the captured volume for irrigation of landscape or natural pervious areas. This technology is very effective as a stormwater quality practice in that, for the captured water quality volume, it provides virtually no discharge to receiving waters and high stormwater constituent removal efficiencies. This technology mimics natural undeveloped watershed conditions wherein the vast majority of the rainfall volume during smaller rainfall events is infiltrated through the soil profile. Their main advantage over other infiltration technologies is the use of an irrigation system to spread the runoff over a larger area for infiltration. This allows them to be used in areas with low permeability soils. Figure 6-1 shows a plan view schematic of a typical holding pond and irrigation area.

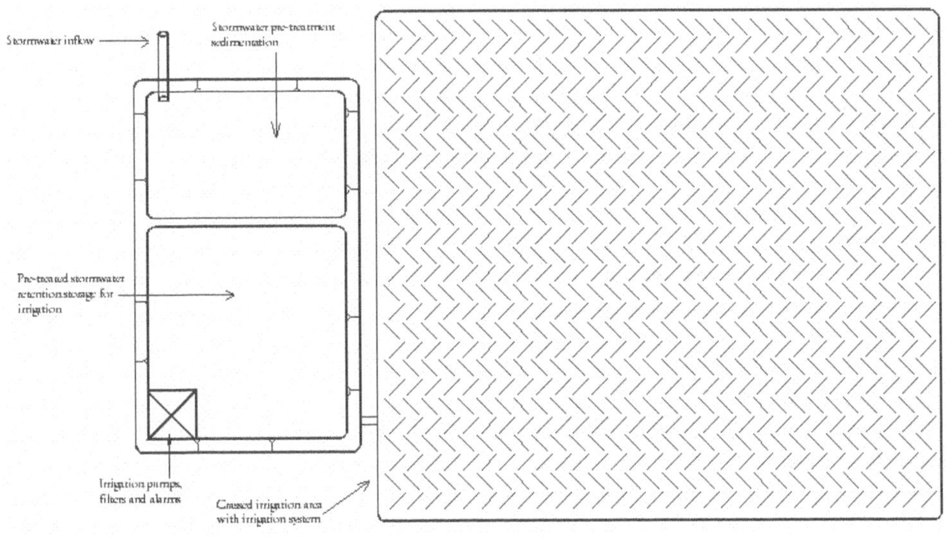

Figure 6-1. Plan View of Holding Pond and Irrigation Area.

Capture of stormwater can be accomplished in almost any kind of runoff storage facility, ranging from dry, concrete-lined ponds to those with vegetated basins and permanent pools. The pump and wet well should be automated with a rainfall sensor to provide irrigation only during periods when required infiltration rates can be realized. Generally, a spray irrigation system is required to provide an adequate flow rate for distributing the water quality volume. Collection of roof runoff for subsequent use (rainwater harvesting) also qualifies as a retention/irrigation practice.

This technology is still in its infancy and there are no published reports on its effectiveness, cost, or operational requirements. The guidelines presented in the remainder of this section should be considered tentative until additional data are available.

6.2 Advantages

- Pollutant removal effectiveness is high, accomplished primarily by (1) sedimentation in the primary storage facility; (2) physical filtration of particulates through the soil profile; and (3) uptake of dissolved constituents in the vegetative root zone by the soil-resident microbial community.
- The hydrologic characteristics of this technique are effective for simulating pre-developed watershed conditions through (1) containment of higher frequency flood volumes (less than about a 2-year event); and (2) reduction of flow rates and velocities for erosive flow events.
- Pollutant removal rates are estimated to be nearly 100% for all pollutants in the captured and irrigated stormwater volume. However, relatively frequent inspection and maintenance is necessary to assure proper operation of these facilities.
- This technology is particularly appropriate for areas with infrequent rainfall because the system is not required to operate often and the ability to provide stormwater for irrigation can reduce demand on surface and groundwater supplies.

6.3 Limitations

- Retention-irrigation is a relatively expensive technology primarily because of mechanical systems, power requirements, and high maintenance needs.
- Due to the relative complexity of irrigation systems, they must be inspected and maintained at regular intervals to ensure reliable system function.
- Retention-irrigation systems use pumps requiring electrical energy inputs (which cost money, create pollution, and can be interrupted). Mechanical systems are also more complex, requiring skilled maintenance, and they are more vulnerable to vandalism than simpler, passive systems.
- Retention-irrigation systems require open space for irrigation and, therefore, may be difficult to retrofit in urban areas.
- Effective use of retention irrigation requires some form of pretreatment of runoff flows (e.g., sediment forebay or vegetated filter) to remove coarse sediment and to protect the long-term operating capacity of the irrigation equipment.

6.4 Design and Sizing Guidelines

The following design guidelines were adopted by the Texas Commission on Environmental Quality (Barrett, 1999).

Table 6-1. Design Guidelines.

Design Criteria	Current Guidance/Recommendations
Storage Facility Sizing	♦ Design of the runoff storage facility is flexible as long as the water quality volume and an appropriate pump and wetwell system can be accommodated.
Pump and Wetwell	♦ A reliable pump, wetwell, and rainfall or soil moisture sensor system should be used to distribute the water quality volume. These systems should be similar to those used for wastewater effluent irrigation, which are commonly used in areas where "no discharge" wastewater treatment plant permits are issued.
Detention Time	♦ The irrigation schedule should allow for complete drawdown of the water quality volume within 72 hours. Irrigation should not begin within 12 hours of the end of rainfall so that direct storm runoff has ceased and soils are not saturated. Consequently, the length of the active irrigation period is 60 hours. ♦ The irrigation should include a cycling factor of ½, so that each portion of the area will be irrigated for only 30 hours during the total of 60 hours allowed for disposal of the water quality volume. Irrigation also should not occur during subsequent rainfall events.
Irrigation System	♦ Generally a spray irrigation system is required to provide an adequate flow rate for timely distribution of the water quality volume.
Irrigation Site Criteria	♦ The area selected for irrigation must be pervious, on slopes of less than 10%. ♦ A geological assessment is required for proposed irrigation areas to assure that there is a minimum of 30 cm (12 in.) of soil cover. Rocky soils are acceptable for irrigation; however, the coarse material (diameter greater than 1.3 cm [0.5 in.]) should not account for more than 30% of the soil volume. ♦ Optimal sites for irrigation include recreational and greenbelt areas as well as landscaping in commercial developments. ♦ The stormwater irrigation area should be distinct and different from any areas used for wastewater effluent irrigation. ♦ The area designated for irrigation should have at least a 100 ft (30 m) buffer from wells, septic systems, and natural wetlands.
Irrigation Area	♦ The irrigation rate must be low enough so that the irrigation does not produce any surface runoff; consequently, the irrigation rate may not exceed the permeability of the soil. The minimum required irrigation area should be calculated using the following formula: $$A = \frac{12 \times V}{T \times r}$$ where: A = area required for irrigation (ft^2) V = water quality volume (ft^3) T = period of active irrigation (30 hr) r = Permeability (in./hr) ♦ The permeability of the soils in the area proposed for irrigation should be determined using a double ring infiltrometer (ASTM D 3385-94) or from county soil surveys prepared by the Natural Resource Conservation Service. If a range of permeabilities is

Design Criteria	Current Guidance/Recommendations
	reported, the average value should be used in the calculation. If no permeability data are available, a value of 0.1 in./hr (0.25 cm/hr) should be assumed. ♦ It should be noted that the minimum area requires intermittent irrigation over a period of 60 hours at low rates to use the entire water quality volume. This intensive irrigation may be harmful to vegetation that is not adapted to long periods of wet conditions. In practice, a much larger irrigation area will provide better use of the retained water and promote a healthy landscape.

6.5 Maintenance Requirements

Table 6-2. Typical Maintenance for Retention/Irrigation System.

Activity	Schedule
♦ The irrigation system should be inspected and tested (or observed while in operation) to ensure proper operation. Two of these inspections should occur during or immediately following wet weather. Any leaks, broken spray heads, or other malfunctions with the irrigation system should be repaired immediately.	Six times annually
♦ Sediment should be removed from inlet structure/sediment forebay, and from around the sump area. When sediment in other areas of the basin fills the 10% volume allocated for sediment accumulation, all sediment should be removed and disposed of properly.	Biannually or when depth reaches 3 in.
♦ The upper stage, side slopes, and embankment of a retention basin must be mowed regularly to discourage woody growth and control weeds. Height of grass in and around basins should be limited to 18 in. More frequent mowing to maintain aesthetic appeal may be necessary in landscaped areas. When mowing is performed, a mulching mower should be used, or grass clippings should be caught and removed.	At least biannually
♦ Debris and litter will accumulate near the basin pump and should be removed during regular mowing operations and inspections. Particular attention should be paid to floating debris that can eventually clog the irrigation system.	As needed
♦ The pond side slopes and embankment may periodically suffer from slumping and erosion, although this should not occur often if the soils are properly compacted during construction. Regrading and revegetation may be required to correct the problems	As needed

6.6 Costs

Overhead and maintenance (O&M) costs for retention-irrigation systems are believed to be high compared to virtually all other stormwater quality control practices because of the need for frequent inspections, the reliance upon mechanical equipment, and power costs.

6.7 Research Needs

This is an emerging BMP and research into almost all phases of design, operation, maintenance, and cost is needed. Important elements include:

- Sizing of the irrigation area;
- Rate at which retained runoff can be applied;
- Construction costs;
- Maintenance frequency for pumps and related irrigation system; and
- Cost of operation.

CHAPTER 7.0

VEGETATED SWALES

7.1 Description

A vegetated swale is a broad, shallow channel with a dense stand of vegetation covering the side slopes and bottom. Swales can be natural or manmade, and are designed to trap particulate pollutants (suspended solids and trace metals), promote infiltration, and reduce the flow velocity of stormwater runoff. A typical example is shown in Figure 7-1, and a schematic is given in Figure 7-2. Vegetated swales can replace curbs, gutters and pipes in a stormwater conveyance system, but are limited to the amount of runoff they can treat effectively. Therefore, swales are best suited for residential, industrial, and commercial areas with low flow and smaller populations.

As stormwater runoff flows through these vegetated channels, sedimentation is promoted by the relatively low velocities and shallow water depths. In addition, some uptake of nutrients by the vegetation may occur and a substantial amount of the runoff infiltrates into the underlying soils. Through geometric modifications and other features, the swale can be designed as both a treatment and conveyance system.

Figure 7-1. Vegetated Swale.

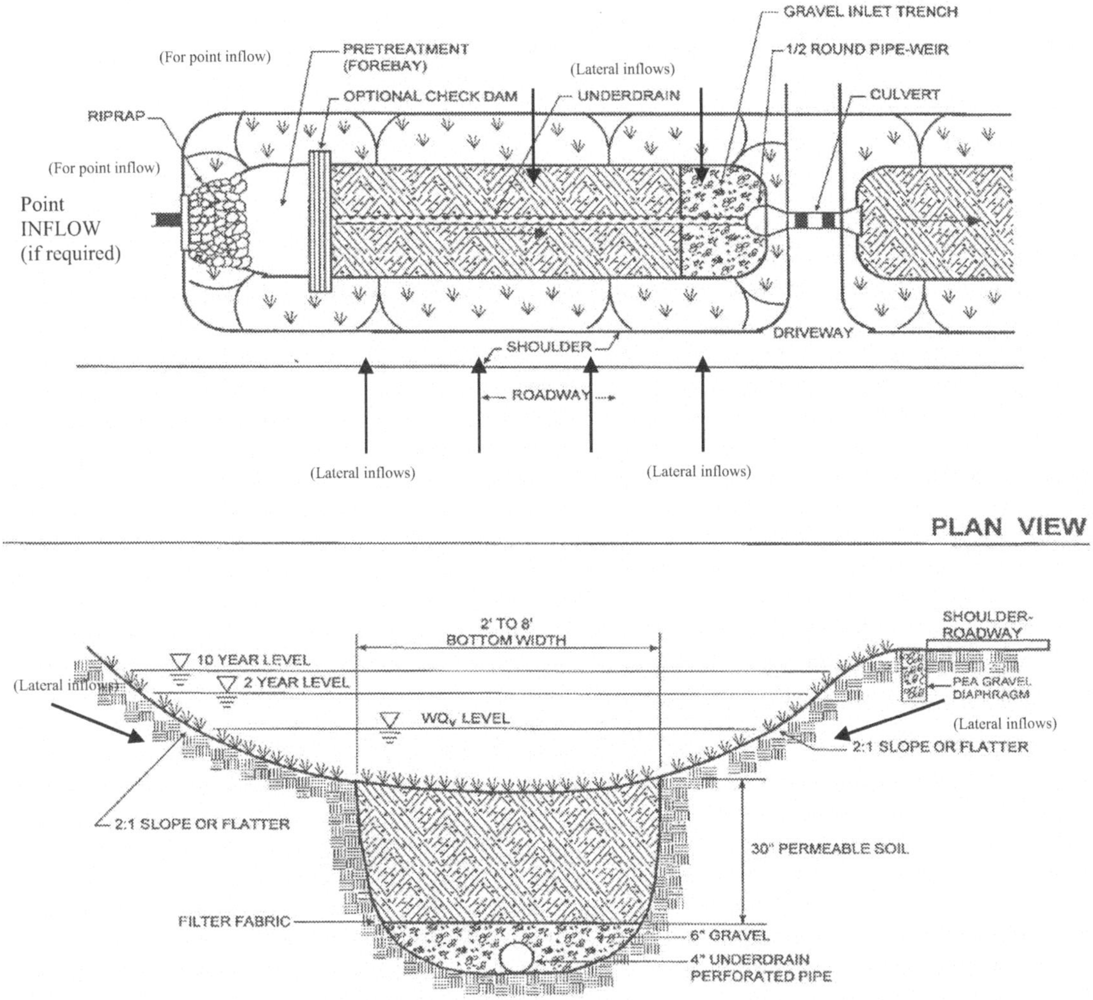

Figure 7-2. Schematic of Swale — U.S. Example (Maryland Department of the Environment, 2000).

In cold or snowy climates, swales may serve a dual purpose by acting as both a snow storage/treatment and a stormwater management practice. This dual purpose is particularly relevant when swales are used to treat road runoff. If used for this purpose, swales should incorporate salt-tolerant vegetation, such as creeping bentgrass.

In arid or semiarid climates, swales should be designed with drought-tolerant vegetation, such as buffalo grass. The value of vegetated practices for water quality management obviously needs to be weighed against the cost of the water needed to maintain them in these conditions.

7.2 Advantages

- If properly designed, vegetated, and operated, swales can serve as an aesthetic, potentially inexpensive urban development or roadway drainage conveyance measure with significant collateral water quality benefits.
- Roadside ditches should be regarded as significant potential swale/buffer strip sites and should be utilized for this purpose whenever possible.

7.3 Limitations

Grassed swales have some limitations, including the following:

- Grassed swales cannot treat a very large drainage area.
- If designed improperly (e.g., if longitudinal slope is too great), grassed channels will have very little pollutant removal.
- A thick vegetative cover is needed for these practices to function properly.
- They are impractical in areas with very flat grades, steep topography, or wet or poorly drained soils.
- They are not effective and may even erode when flow volumes and/or velocities are high.
- Sufficient land may not be available for suitable swale designs to be incorporated.
- In some places in the United States, their use is restricted by law. Local municipalities/councils in the United States and the United Kingdom may require curb and gutter systems in residential areas.
- They are impractical in areas with erosive soils or where a dense vegetative cover is difficult to maintain.
- Infiltration through the swale may carry pollutants into local groundwater.
- Standing water in vegetated swales can lead to potential safety, odor, and mosquito problems.

7.4 Siting Criteria

The suitability of a swale at a site depends on land use, size of the area serviced, soil type, slope, imperviousness of the contributing watershed, and dimensions and slope of the swale system (Schueler et al., 1992). In general, swales can be used to serve areas of less than 10 ac, with slopes no greater than 6%. The seasonal high water table should be at least 4 ft below the surface. Use of natural topographic lows is encouraged and historic natural drainage courses should be considered, where appropriate (Young et al., 1996). Roadside ditches should be regarded as significant potential swale/buffer strip sites and should be utilized for this purpose whenever possible. However, the use of swales in areas subject to frequent roadside parking should be avoided, as this can cause over-compaction of the soil and deterioration of the surface (CIRIA, 2000a,b).

Research in the Austin, Texas, area indicates that vegetated controls are effective at removing pollutants even when dormant (Barrett et al., 1998). Therefore, irrigation is not required to maintain growth during dry periods, but may be necessary only to prevent the vegetation from dying.

The topography of the site should permit the design of a channel with appropriate slope and cross-sectional area. Site topography may also dictate a need for additional structural controls. Recommendations for longitudinal slopes range between 2% and 6%. Flatter slopes can be used, if sufficient to provide adequate conveyance. Steep slopes increase flow velocity, decrease detention time, and may require energy dissipating and grade checks. Steep slopes also can be managed using a series of check dams to terrace the swale and reduce the slope to within acceptable limits. The use of check dams with swales also promotes infiltration and increases flow attenuation.

7.5 Design Guidelines

Swales can be used purely as a conveyance system to direct and convey runoff from the drained area to another stage of the surface water management system. Swales can also be designed for runoff attenuation, for treatment, and for disposal (using infiltration through the base of the swale) (CIRIA, 2000a,b).

Almost all of the current guidelines for swales both in the United States and the United Kingdom are founded on the results of a single study conducted in the United States (Seattle Metro and Washington Department of Ecology, 1992). This study is the basis for hydraulic residence time requirements, appropriate slopes, and maximum velocities for the water quality design storm. Most studies conducted subsequently were on swales designed according to these recommendations and generally confirm that substantial pollutant removal does occur in the recommended design, but do little to expand on whether similar performance might be observed in other configurations.

In the United States, a trapezoidal cross-section is normally recommended (see, for example, WSDOT, 1995); however, the main advantage seems to be the ease of calculating water depth for the design storm intensity. Barrett et al. (1998) demonstrated that when runoff enters the swale from the side, the majority of the pollutant removal occurs on the side slopes; consequently, other geometries, including wider, flatter sides may actually perform better.

Swales to be used for conveyance should be designed in accordance with standard hydraulic principles. Commonly recommended design guidelines are summarized in Table 7-1.

Table 7-1. Design Guidelines.

Design Criteria	Current Guidance/Recommendations
Hydraulic Residence Time (Swale Length)	♦ Most of the design guidelines adopted for swale design specify a minimum hydraulic residence time of 9 minutes. This criterion is based on the results of a single study conducted in Seattle, Washington (Seattle Metro and Washington Department of Ecology, 1992), and is not well supported. Analysis of the data collected in that study indicates that pollutant removal at a residence time of 5 minutes was not significantly different, although there is more variability in that data.
Planting	♦ Swales must be vegetated to provide adequate treatment of runoff. It is important to maximize water contact with vegetation and the soil surface. ♦ Pollutant removal efficiencies are not very dependent on the specific plants involved, so the vegetation should be selected with stabilization of the soils as the primary consideration (Barrett et al., 1998; Caltrans, 2002). ♦ A fine, close-growing, water-resistant grass should be selected for use in vegetated swales. Guides to swale design often suggest a grass length; for example, 150 mm (0.5 ft), suggested in SEPA (1999). Many design guidelines recommend that grass be mowed frequently to maintain dense coverage near the ground surface, and observational experience would suggest that short-cut grass is most likely to be accepted in the United Kingdom. Recent research (Colwell et al., 2000), however, has shown mowing frequency or grass height has little or no effect on pollutant removal. ♦ The vegetation in the base of the swale should be maintained at or just above the maximum water level for the water quality storm (CIRIA, 2000a,b).
Side Slopes	♦ The side slopes should be no steeper than 3:1 (H:V) (WSDOT, 1995). ♦ The sides of the swale should not be steeper than 1 in 4, to allow for mowing and access for maintenance personnel.
Bottom Width	♦ The maximum bottom width is 2.6 m (10 ft) unless a dividing berm is provided (WSDOT, 1995). This provision will discourage formation of gullies and ensure there is a consistent spread of flow across the swale.
Flow Depths	♦ The depth of flow should not exceed 10 cm (4 in.) during a 2.5 cm/hr (1 in./hr) storm. ♦ The depth of flow should not exceed 0.1 m (0.33 ft) (CIRIA, 2000a,b) ♦ The flow depth should be less than the height of the grass to ensure filtration (CIRIA, 2000a,b) or a grass swale's depth should not exceed one-third of the grass height in infrequently mowed swales, or one half of the grass height in regularly mowed swales (Horner et al., 1994)
Longitudinal Channel Slope	♦ The channel slope should be at least 1% (in wet climates) and no greater than 5% (WSDOT, 1995), although recent work (Colwell, 2001) suggests that stability is increased if the slope does not exceed 2.5%. ♦ The longitudinal bottom slope of all types of swale should be kept as level as possible, and ideally no greater than 2% (CIRIA, 2000a,b) ♦ Steeper overall longitudinal slopes can be accommodated by the introduction of check dams spaced at intervals along the length of the swale. Criteria for designing the check dams are as follows (CIRIA, 2000a,b): ♦ They should be designed to retain the runoff upstream. ♦ Interconnections between adjacent ponds should be designed so that the flow between them does not resuspend settled material or cause local erosion. ♦ Interconnections between adjacent sections should be designed to retain floating solids or surface films. ♦ Where slopes are greater than 6%, infiltration will not be effective and the system will only act as a conveyance device (CIRIA, 2000a,b)

Design Criteria	Current Guidance/Recommendations
Flow Velocities	♦ Flow velocities for side-slope flows should be less than 0.3 m/s (1 ft/s) in order to promote filtration and settlement (CIRIA, 2000a,b). ♦ The runoff velocity in the swale should not exceed 0.3 m/s (1 ft/s) at the design treatment volume (CIRIA, 2000a,b). ♦ At full conveyance depth, the velocity should not exceed 1.5m/s (CIRIA, 2000a,b).
Inlet Infrastructure	♦ If flow is to be introduced through curb cuts, place pavement slightly above the elevation of the vegetated areas. Curb cuts should be at least 12 in. (30 cm) wide to prevent clogging (WSDOT, 1995). ♦ Inflow could be via a low earth weir at the edge of the swale (CIRIA, 2000a,b). ♦ Inflow could be via a riprap or concrete chute, although in this case erosion protection measures will be required to protect the surface of the swale (CIRIA, 2000a,b). ♦ Any inlet infrastructure that directs the flow to the base of the swale bypasses the beneficial filtering effect of the side slopes.
Pretreatment	♦ Pretreatment is required to protect the swale. For pipe inlets, 0.1 in. (0.25 cm) per contributing acre should be temporally stored behind a check dam. For lateral inflows gentle slope or a pea gravel diaphragm can be used (Schueler, 1997a).
Subsoil	♦ It is necessary to ensure that the design soil characteristics adequately represent conditions across the full length of the swale. Where additional ground investigation is not available, trial pits should be dug at 25 m (82 ft) intervals (CIRIA, 2000a,b). ♦ It is often necessary to modify the parent soils to improve the infiltration rate. Dry swales will have a prepared soil filter bed that is composed of 50% sand, 50% silt loam. Swale filter beds are drained by a longitudinal perforated pipe to keep swale dry after storm events (Schueler, 1997a). ♦ The underdrain system should be composed of a 6 in. (15 cm) gravel bed with a 4 in. (10 cm) polyvinylchloride (PVC) pipe. ♦ The soils used to finish the swale slopes should be suitably fertile, porous and of sufficient depth to ensure healthy vegetation growth (CIRIA, 2000a,b).
Water Table	♦ Geotechnical tests must be performed to determine the location of the water table. If the water table is within 2 ft (0.6 m) of the planned swale bottom, a dry swale is not feasible (Schueler, 1997a). ♦ If there is a risk of a swale conveying potential polluted water, and the groundwater requires protection, an impermeable liner should be incorporated in the design (CIRIA, 2000a,b).
Shape	♦ The base of the swale should be flat, with a smooth transition between the base and the sides (CIRIA, 2000a,b).

7.6 Construction Guidelines

- If possible, divert runoff (other than necessary irrigation) during the period of vegetation establishment. Where runoff diversion is not possible, cover graded and seeded areas with suitable erosion control materials.
- The bottom and side slopes of the swale should be carefully prepared to ensure that they are structurally sound, level, and true and will perform their required hydraulic function. They should not be compacted during construction (CIRIA, 2000a,b).

7.7 Performance

The literature suggests that vegetated swales represent a practical and potentially effective technique for controlling urban runoff quality. It is known that check dams, slight slopes, permeable soils, dense grass cover, increased contact time, and small storm events all contribute to successful pollutant removal by the swale system. Factors decreasing the effectiveness of swales include compacted soils, short runoff contact time, large storm events, frozen ground, short grass heights, steep slopes, and high runoff velocities and discharge rates.

Vegetated swales and filter strips have not been accepted widely as primary controls for the treatment of stormwater runoff. This is mainly the result of the wide range of pollutant removals reported for vegetative controls in various studies (Schueler et al., 1992; Young et al., 1996; U.S. EPA, 1983). Consequently, a lack of confidence emerged among regulatory agencies that vegetative controls could provide reliable and consistent removal of pollutants in stormwater. Most design manuals recommend vegetative controls only for pretreatment to reduce sediment loading to filtration systems or other structural stormwater controls. Unfortunately, many of the studies in which lower removal efficiencies were observed were not well designed and significant removal of the pollutants occurred before the runoff entered the test sections that were monitored.

One example is the evaluation of a grassy swale in Austin, Texas, by Welborn and Veenhuis (1987). The authors reported low or negative removals for many stormwater constituents; however, the runoff had traveled through a grassy swale for more than 60 m (200 ft) *before* reaching the influent to the site that was monitored. The median influent suspended solids concentration was less than 20 mg/L, even though the drainage was derived from an area of medium density townhouses. Unpublished data collected by the City of Austin indicate that TSS concentrations in stormwater runoff from this type of land use are typically above 100 mg/L. The low influent concentration indicates that most of the removal occurred before entering the test section; therefore, the removals reported understate the potential improvement in water quality resulting from use of vegetative controls.

Kaighn and Yu (1996) recognized that the quality of highway runoff entering two test swales was considerably better than that observed at an edge of pavement site located nearby. For example, average TSS concentrations in runoff entering the swale were 38.7 and 32.8 mg/L, while runoff sampled from the pavement nearby had a TSS concentration of 112.9 mg/L. Additional monitoring indicated that much of the removal occurred in the vegetated filter strip crossed by the runoff before entering the swale test section.

Dorman et al. (1996) analyzed the performance of three vegetated channels for treating highway runoff and reported TSS removal efficiencies ranging from 98% to -7%. Operational problems and erosion of the channel were reported at the Virginia site. The influent TSS concentrations were as low as 8 mg/L at the Maryland site, and the average was less than 30 mg/L. The third site in Florida was extremely effective in removing TSS. The wide range of reported removal efficiencies reinforces the belief that small changes in channel characteristics cause large changes in pollutant removal effectiveness; however, only the data from the Florida site are indicative of the pollutant removal that might be expected in grassy channels.

The stormwater discharged from the vegetated controls in many of these studies contained 20 mg/L or less of TSS, which is well below the concentrations that would be expected based on the land use in the watershed. However, the impression conveyed by these published reports is that the vegetative controls are unreliable, or worse do nothing at all.

In the studies described above, the monitored vegetative controls generally operated to polish the quality of the discharge rather than as the primary treatment device. A more efficient and consistent performance would be expected when the concentrations in the influent to the controls are similar to what might be encountered in untreated urban or highway runoff. The TSS concentrations might range from 100 to 200 mg/L, which is almost an order of magnitude higher than the influent concentrations reported in the cited studies.

Although TSS removal may be substantial, performance for other constituents may be less impressive. A project in Durham, North Carolina, monitored performance of a carefully designed artificial swale receiving runoff from a commercial parking lot. The project tracked 11 storms and concluded that particulate concentrations of heavy metals (Cu, Pb, Zn, and Cd) were cut by approximately 50%, although the swale was largely ineffective at removing soluble nutrients.

Caltrans (2002) conducted a study of the efficiency of six swales located in Southern California, and the detailed results of this study are presented in Table 7-2. The column titled "Significance" indicates the probability that the mean influent and effluent concentrations are not significantly different, based on an ANOVA. At the test sites, approximately 50% of the runoff infiltrated, so the total load reduction is much greater than the concentration reduction shown in Table 7-2.

Table 7-2. Performance of Vegetated Swales (Caltrans, 2002).

Constituent	Mean EMC		Removal %	Significance P
	Influent mg/L	Effluent mg/L		
TSS	94	47	49	.002
NO_3-N	1.22	0.89	27	.147
TKN	3.43	2.36	31	.907
Total N[a]	4.64	3.24	30	-
Orthophosphate	0.13	0.40	-218	<.000
Phosphorus	0.26	0.53	-106	.001
Total Cu	0.049	0.019	63	<.000
Total Pb	0.099	0.031	68	.075
Total Zn	0.349	0.079	77	<.000
Dissolved Cu	0.024	0.012	49	.067
Dissolved Pb	0.018	0.007	57	.081
Dissolved Zn	0.170	0.045	74	<.000
TPH-Oil [b]	3.5	1.7	51	.107
TPH-Gasoline [b]	<0.05[c]	<0.05[c]	NA	-
TPH-Diesel [b]	1.3	0.4	69	.156
Fecal Coliform [b]	12,300 MPN/100mL	16,000 MPN/100mL	-30	.707

[a] Considered to be sum of NO_3-N and TKN-N.
[b] TPH and coliform are collected by grab method and may not accurately reflect removal.
[c] Equals value of reporting limit.

A summary of the results of other studies that have been conducted on a variety of grassed channels is presented in Table 7-3.

Table 7-3. Grassed Swale Pollutant Removal Efficiency Data.

	Removal Efficiencies (% Removal)						
Study	**TSS**	**TP**	**TN**	**NO_3**	**Metals**	**Bacteria**	**Type**
Caltrans, 2002	77	8	67	66	83-90	-33	
Goldberg, 1993	67.8	4.5	-	31.4	42-62	-100	grassed channel
Seattle Metro and Washington Department of Ecology, 1992	60	45	-	-25	2-16	-25	grassed channel
Seattle Metro and Washington Department of Ecology, 1992	83	29	-	-25	46-73	-25	grassed channel
Wang et al., 1981	80	-	-	-	70-80	-	dry swale
Dorman et al., 1989	98	18	-	45	37-81	-	dry swale
Harper, 1988	87	83	84	80	88-90	-	dry swale
Kercher, Landon, and Massarelli, 1983	99	99	99	99	99	-	dry swale
Harper, 1988.	81	17	40	52	37-69	-	wet swale
Koon, 1995	67	39	-	9	-35 to 6	-	wet swale
Occoquan Watershed Monitoring Lab, 1983	-100	-100	-100	-	-100	-	drainage channel
Yousef et al., 1985	-	8	13	11	14-29	-	drainage channel
Occoquan Watershed Monitoring Lab, 1983	-50	-9.1	-18.2	-	-100	-	drainage channel
Yousef et al., 1985	-	-19.5	8	2	41-90	-	drainage channel
Occoquan Watershed Monitoring Lab, 1983	31	-23	36.5	-	-100 to 33	-	drainage channel
Welborn and Veenhuis, 1987	0	-25	-25	-25	0	-	drainage channel
Yu and Gerde, 1993	68	60	-	-	74	-	drainage channel
Dorman et al., 1989	65	41	-	11	14-55	-	drainage channel
Pitt and McLean, 1986	0	-	0	-	0	0	drainage channel
Oakland, 1983	33	-25	-	-	20-58	0	drainage channel
Dorman et al., 1989	-85	12	-	-100	14-88	-	drainage channel
Lawrence et al., 1997	20-40	20-40	20-40	-	0-20	20-40	dry swale
Roesner et al., 1997	80	40	40	40	50-75	-	dry swale
Honeyman and Baird, 2001	80						dry swale
Honeyman and Baird, 2001	44						dry swale
Macdonald and Jefferies, 2001	72		100				dry swale
Macdonald and Jefferies, 2001	55	33	-42		43 –423??		dry swale

Removal Efficiencies (% Removal)							
Study	TSS	TP	TN	NO_3	Metals	Bacteria	Type
Backstrom, 2001	79-98						Laboratory experiment
Roesner et al., 1997	80	40	40	40	50-75	-	dry swale

The data confirm the relatively high removal rates for some pollutants, but negative removals for bacteria, and fair performance for nutrients. Field performance data have been gathered for two swales in Scotland and the results are currently being written up (Macdonald, 2002).

Although it is difficult to distinguish between different designs based on the small amount of available data, grassed channels generally have poorer removal rates than wet and dry swales, although some swales appear to export soluble phosphorus (Harper, 1988; Koon, 1995). It is not clear why swales export bacteria. One explanation is that bacteria thrive in the warm swale soils.

Researchers in Sweden have tested the performance of laboratory constructed swales and working swales using a standardized runoff event simulation procedure. Backstrom's (2001) studies focused on the removal of suspended solids and demonstrated an average of 79% to 98% removal under different vegetative conditions.

Figures 7-3 through 7-7 provide a summary of the tabulated results (Table 7-3). From the results, it appears that standard drainage channels have a much more variable performance than designed swales, and their removal efficiencies are, therefore, not analyzed further.

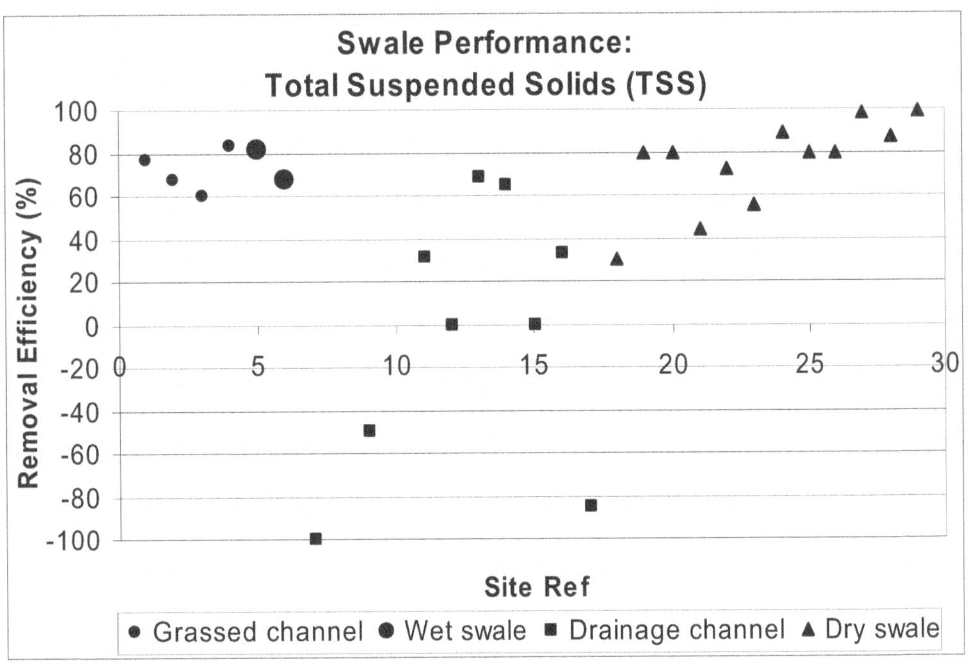

Figure 7-3. Swale Performance for Removal of TSS.

Mean performance levels (in terms of efficiency of removal of total suspended solids) are as follows:

- Grassed channels: 72%
- Wet swales: 74%
- Dry swales: 74%

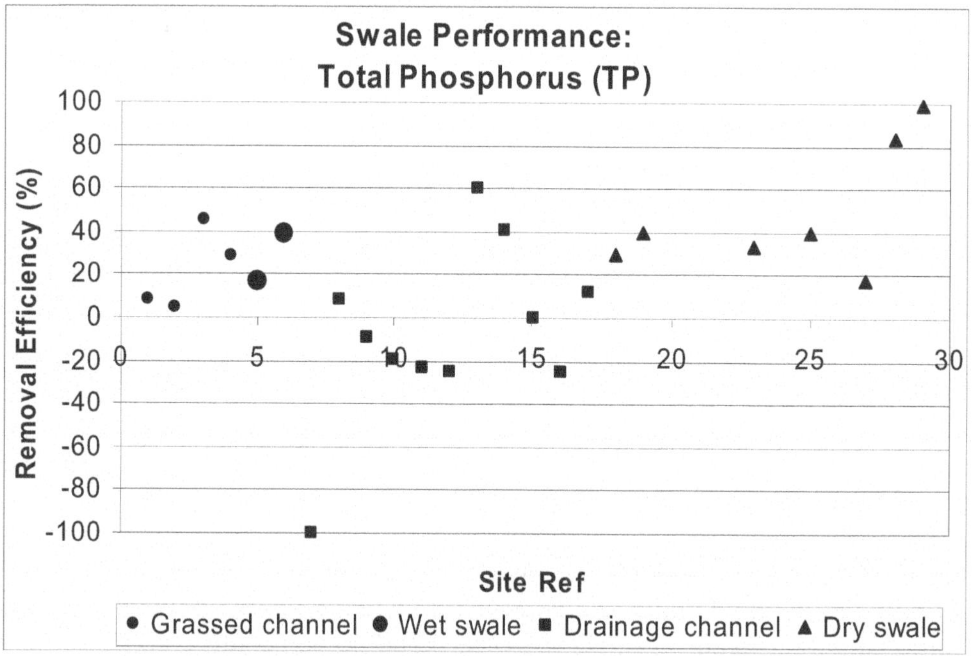

Figure 7-4. Swale Performance for Removal of TP.

Mean performance levels (in terms of efficiency of removal of total phosphorus) are as follows:

- Grassed channels: 22%
- Wet swales: 28%
- Dry swales: 49%

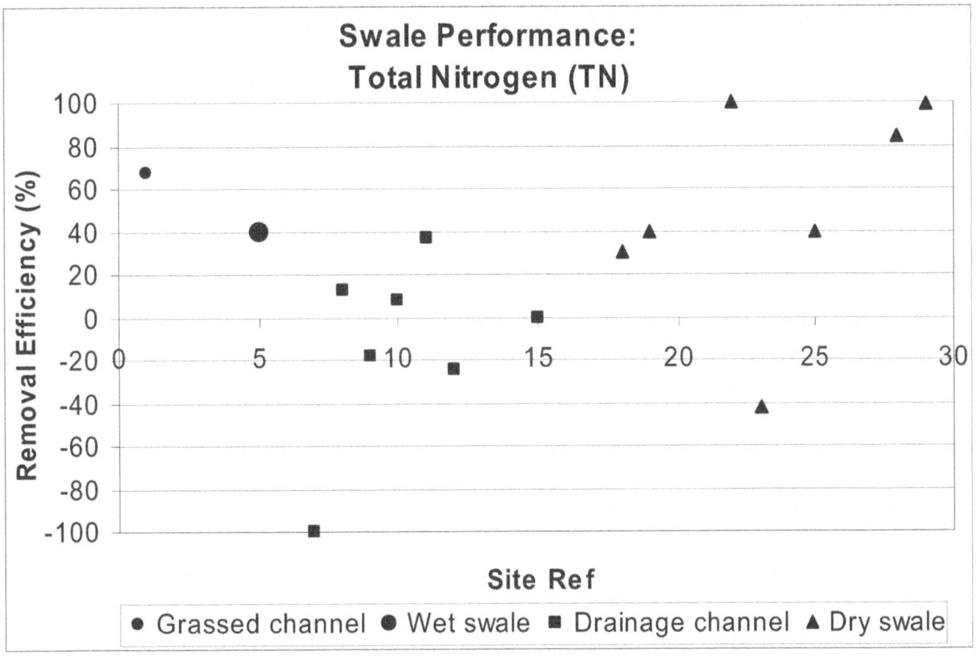

Figure 7-5. Swale Performance for Removal of TN.

Mean performance levels (in terms of efficiency of removal of total nitrogen) are as follows:

- Grassed channels: 67% (based on one observation only)
- Wet swales: 40% (based on one observation only)
- Dry swales: 50%

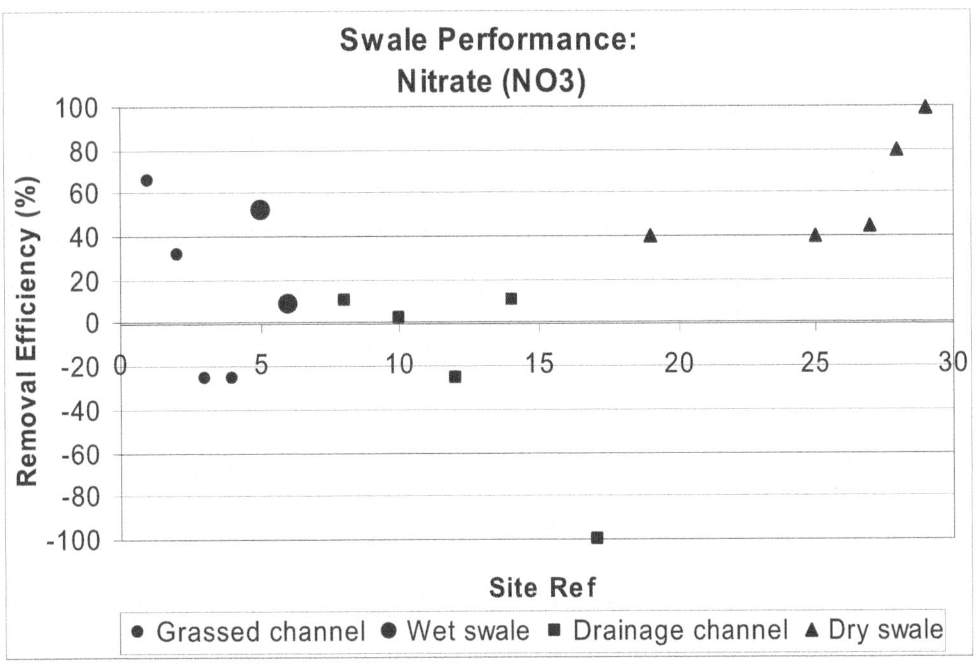

Figure 7-6. Swale Performance for Removal of NO_3.

Mean performance levels (in terms of efficiency of removal of nitrites) are as follows:

- Grassed channels: 12%
- Wet swales: 31%
- Dry swales: 61%

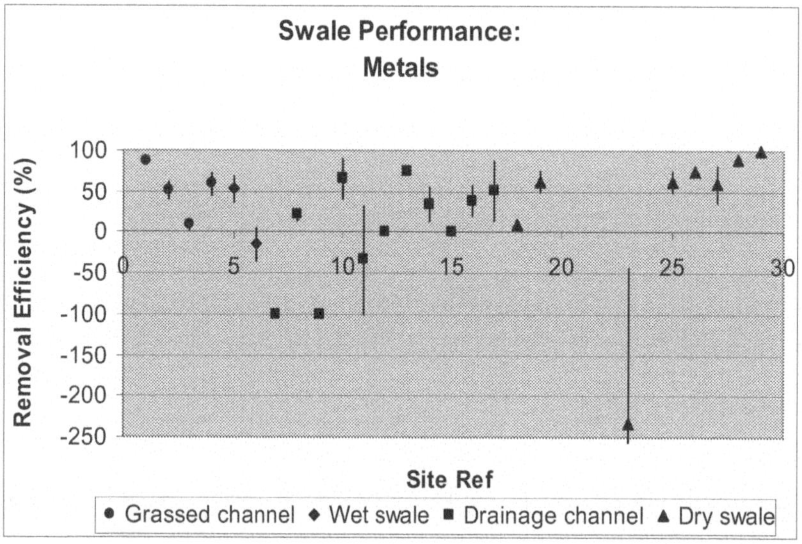

Figure 7-7. Swale Performance for Removal of Heavy Metals.

Figure 7-7 shows high levels of variability insofar as heavy metal removals across all swale types.

The effectiveness of vegetated swales can be enhanced by adding check dams at approximately 17 m (50 ft) increments along their length (see Figure 7-1). These dams maximize the retention time within the swale, decrease flow velocities, and promote particulate settling. Finally, the incorporation of vegetated filter strips parallel to the top of the channel banks can help to treat sheet flows entering the swale.

7.8 Maintenance

The useful life of a vegetated swale system is directly proportional to its maintenance frequency. The maintenance objectives for vegetated swale systems include keeping up the hydraulic and removal efficiency of the channel and maintaining a dense, healthy grass cover.

Maintenance activities should include periodic mowing (with grass never cut shorter than the design flow depth); weed control; watering during drought conditions, if needed to keep the vegetation alive; reseeding of bare areas; and clearing of debris and blockages. The frequency of mowing will be largely dependent on the species of grass planted but is likely to be necessary at least two to three times per year. Mowing and other maintenance equipment should not damage or excessively consolidate the surface (CIRIA, 2000a,b). Cuttings should be removed from the channel and disposed in a local composting facility. Accumulated sediment should also be removed manually to avoid concentrated flows in the swale. The application of fertilizers and pesticides should be minimized.

Another aspect of a good maintenance plan is repairing damaged areas within a channel. For example, if the channel develops ruts or holes, it should be repaired utilizing a suitable soil that is properly tamped and seeded. The grass cover should be thick; if it is not, reseed as necessary. In the United Kingdom, it is recommended that any standing water removed during the maintenance operation be disposed of to a sanitary sewer at an approved discharge location. Residuals (e.g., silt, grass cuttings) must be disposed in accordance with local or state requirements. Typical maintenance activities are summarized in Table 7-4.

Table 7-4. Typical Maintenance Activities for Grassed Swales (Source: Adapted from CWP, 1996).

Activity	Schedule
♦ Inspect pea gravel diaphragm for clogging and correct the problem. ♦ Inspect grass along side slopes for erosion and formation of rills or gullies and correct. ♦ Remove trash and debris accumulated in the inflow forebay. ♦ Inspect and correct erosion problems in the sand/soil bed of dry swales. ♦ Based on inspection, plant an alternative grass species, if the original grass cover has not been successfully established. ♦ Replant wetland species (for wet swale), if not sufficiently established.	Annually (semi-annually the first year)
♦ Rototill or cultivate the surface of the sand/soil bed of dry swales, if the swale does not draw down within 48 hours. ♦ Remove sediment build-up within the bottom of the swale once it has accumulated to 25% of the original design volume and scarify to encourage vegetation redevelopment or replant as necessary.	As needed (infrequent)
♦ Mow grass as needed for safety, aesthetic, or other purposes.	As needed

7.9 Design Lives

- If properly designed and regularly maintained, vegetated swales can last indefinitely (U.S. EPA, 1999c).
- 10-year effective life (North Carolina Department of Environment and Natural Resources, undated).
- 5-20 year effective life (FWHA, 1999).

7.10 Costs

7.10.1 Construction Costs

Limited data are available for estimating the difference in cost between various swale designs. One study (SWRPC, 1991) estimated the construction cost of grassed channels at approximately $2.70/m^3 ($0.25 per ft^2). This price does not include design costs or contingencies. Brown and Schueler (1997) estimate these costs at approximately 32% of construction costs for most stormwater management practices. For swales, however, these costs would probably be significantly higher because the construction costs are so low compared with other practices. A more realistic estimate would be a total cost of approximately $0.5.40/m^3 ($0.50 per ft^2), which compares favorably with other stormwater management practices.

Table 7-5 presents the construction costs compiled by Caltrans (2001) from a number of locations across the United States, whereas 7-6 presents unit costs estimated by SWRPC (1991). No data are currently available in the United Kingdom regarding construction cost of swales, although a figure of £5–£10/m has been cited.

Table 7-5. U.S. Swale Construction Costs (Caltrans, 2001).

Entity	Distance Area (acres)	Water Quality Volume (ft^3)	Adjusted Total Cost	Adjusted Total Cost per Acre Treated
Maryland and Virginia CWP	80.00	-	$ 3561	$ 45
Maryland and Virginia CWP	5.00	-	$ 18,932	$ 3786
Maryland and Virginia CWP	0.87	6,778	$ 18,089	$ 20,792
ODOT	1.17	-	$ 41,736	$ 35,672
Caltrans	2.40	6,916	$ 136,822	$ 57,009
Caltrans	2.30	6,650	$ 140,006	$ 60,872
Caltrans	0.40	1,742	$ 31,992	$ 79,979
Caltrans	0.70	2,614	$ 70,138	$ 100,197
Caltrans	0.70	2,178	$ 76,179	$ 108,827
Caltrans	0.50	1,742	$ 125,488	$ 250,977

Table 7-6. Swale Cost Estimate (SWRPC, 1991).

Component	Unit	Extent	Unit Cost			Total Cost		
			Low	Moderate	High	Low	Moderate	High
Mobilization/Demobilization-Light	Swale	1	\$107	\$274	\$441	\$107	\$274	\$441
Site Preparation								
Clearing[b]								
Grubbing[c]	Acre	0.5	\$2200	\$3800	\$5400	\$1100	\$1900	\$2700
General	Acre	0.25	\$3800	\$5200	\$6600	\$950	\$1300	\$1650
Excavation[d]	Yd^3	372	\$2.10	\$3.70	\$5.30	\$781	\$1376	\$1972
Level and Till[e]	Yd^2	1,210	\$0.20	\$0.35	\$0.50	\$242	\$424	\$605
Sites Development								
Salvaged Topsoil								
Seed and Mulch[f]	Yd^2	1,210	\$0.40	\$1.00	\$1.60	\$484	\$1,210	\$1,936
Sod[g]	Yd^2	1,210	\$1.20	\$2.40	\$3.60	\$1,452	\$2,904	\$4,356
Subtotal	—	—	—	—	—	\$5,116	\$9,388	\$13,660
Contingencies	Swale	1	25%	25%	25%	\$1,279	\$2,347	\$3,415
Total	—	—	—	—	—	\$6,395	\$11,735	\$17,075

Source: (SWRPC, 1991)

Note: Mobilization/demobilization refers to the organization and planning involved in establishing a vegetative swale.
[a] Swale has a bottom width of 1.0 foot, a top width of 10 feet with 1:3 side slopes, and a 1000 foot length.
[b] Area cleared = (top width + 10 feet) x swale length.
[c] Area grubbed = (top width x swale length).
[d] Volume excavated = (0.67 x top width x swale depth) x swale length (parabolic cross-section).

[e] Area tilled = (top width + $\frac{8(\text{swale depth}^2)}{3(\text{top width})}$ x swale length (parabolic cross-section).

[F] Area needed = area cleared x 0.5.
[g] = Area sodded = area cleared x 0.5.

7.10.2 Maintenance Costs

Table 7-7 presents the expected annual maintenance cost for a swale with a tributary area of approximately 2 ha (5 ac) as estimated by Caltrans (2002). Essentially, all the activities are related to vegetation management; consequently, no special training is required for maintenance personnel.

Table 7-7. Expected Annual Maintenance Costs for Swales.

Activity	Labor Hours	Equipment & Material (\$)	Cost
Inspections	1	0	44
Maintenance	47	182	2,250
Vector Control	0	0	0
Administration	3	0	132
Materials	-	310	310
Total	**51**	**\$492**	**\$2,736**

Estimated maintenance costs for two sizes of swales, as presented by SWRPC (1991), are presented in Table 7-8. Annual cost for maintaining vegetated swales is approximately \$1.90 per linear meter for a 0.5 m (1.5 ft) deep channel. Barrett (1999) suggests \$16/m (\$5/ft) is appropriate.

Limited data are available in the United Kingdom regarding maintenance cost of swales, although £35 per tributary acre per year has been cited.

Table 7-8. Estimated Maintenance Costs (SWRPC, 1991).

Component	Unit Cost	Swale Size (Depth and Top Width)		Comment
		1.5 Foot Depth, One-Foot Bottom Width, 10-Foot Top Width	3-Foot Depth, 3-Foot Bottom Width, 21-Foot Top Width	
Lawn Mowing	\$0.85 / 1,000 ft^2/ mowing	\$0.14 / linear foot	\$0.21 / linear foot	Lawn maintenance area=(top width + 10 feet) x length. Mow eight times per year
General Lawn Care	\$9.00 / 1,000 ft^2/ year	\$0.18 / linear foot	\$0.28 / linear foot	Lawn maintenance area = (top width + 10 feet) x length
Swale Debris and Litter Removal	\$0.10 / linear foot / year	\$0.10 / linear foot	\$0.10 / linear foot	--
Grass Reseeding with Mulch and Fertilizer	\$0.30 / yd^2	\$0.01 / linear foot	\$0.01 / linear foot	Area revegetated equals 1% of lawn maintenance area per year
Program Administration and Swale Inspection	\$0.15 / linear foot / year, plus \$25 / inspection	\$0.15 / linear foot	\$0.15 / linear foot	Inspect four times per year
Total	--	**\$0.58 / linear foot**	**\$ 0.75 / linear foot**	--

7.11 Research Needs

7.11.1 Siting Criteria

Siting criteria for swales are well established, and no additional work in this area is recommended.

7.11.2 Design Guidelines

There are a number of uncertainties in the specific effect of several commonly adopted design guidelines, including residence time and side slopes. Clearly, a longer residence time allows more of the finer particles to settle out; however, there is likely an irreducible minimum concentration effluent concentration that can be obtained. Many requirements are based on the need to maintain channel stability during larger storms; consequently evaluation of minor differences in slope or length is a low priority. As with many other devices, a hydrological analysis of the potential changes in storm frequency and intensity would be useful for establishing the long-term viability of current designs.

7.11.3 Performance

Uncertainties about performance include:

- Best designs for optimal pollutant removal capability;
- Relationship of removal rates to age;
- Benefits of check dams or steps in preventing downstream migration of pollutants; and
- Effects of different grass types and height on pollutant removal capability.

7.11.4 Construction Guidelines and Cost

The wide range of cost data reported in Table 7-5 indicates that much more information is needed in this area. Do the differences in cost reflect different regulatory requirements, differences in construction techniques, or perhaps site suitability? Identifying the primary elements that drive construction costs upward is a high priority and could help reduce the initial cost associated with this widely used BMP/SUDS.

7.11.5 Maintenance Guidelines and Cost

Maintenance guidelines are one area in which common sense type recommendations are often derived from a common source that may, in fact, have little empirical basis. Many recommendations seem to derive from aesthetic considerations and may have little impact on performance. The only study for which a detailed accounting of maintenance activities was compiled was conducted in a Mediterranean climate by Caltrans and likely is not representative of the frequency or cost of maintenance in other areas. The development of empirical data related to required maintenance activities and their frequency and cost is a high priority.

7.12 Additional Resources

Reeves, E., 1994, Performance and Condition of Biofilters in the Pacific Northwest. *Watershed Protection Techniques* 1(3):117–119.

U.S. Environmental Protection Agency (U.S. EPA), 1993, *Guidance Specifying Management Measures for Sources of Nonpoint Pollution in Coastal Waters*. EPA-840-B-92-002. U.S. Environmental Protection Agency, Office of Water, Washington, D.C.

CHAPTER 8.0

VEGETATED BUFFER STRIPS

8.1 Description

Grassed buffer strips (vegetated filter strips, filter strips, and grassed filters) are vegetated surfaces that are designed to treat sheet flow from adjacent surfaces. Filter strips function by slowing runoff velocities and allowing sediment and other pollutants to settle, and by providing some infiltration into underlying soils. Buffer strips were originally used as an agricultural treatment practice, and have more recently evolved into an urban practice. With proper design and maintenance, buffer strips can provide relatively high pollutant removal. In addition, the public views them as landscaped amenities and not as stormwater infrastructure. Consequently, there is little resistance to their use.

Buffer strips can be divided into two basic types of vegetative stormwater quality management practices: natural buffers and engineered biofilters. Both systems encourage diffuse, shallow, low-velocity flow across a vegetated surface and both can provide effective stormwater quality enhancement if flow and vegetative conditions are correct. Natural vegetated buffers are typically left undisturbed and are used in areas with relatively low-density development as a passive, low maintenance means of protecting nearby receiving waters from marginally increased pollutant loads.

Engineered biofilters are specifically designed and constructed to maximize the water quality benefits of vegetative filtration practices, particularly in areas where adequate buffers do not exist naturally or can not be preserved. Engineered systems feature mechanically graded surfaces with vegetation selection targeted to physical site conditions, water quality performance, and aesthetic appeal. Figure 8-1 presents a schematic representation of an engineered biofilter; Figure 8-2 shows a photograph of an actual installation.

8.2 Advantages

- Buffers require minimal maintenance (generally just erosion prevention and mowing).
- If properly designed, vegetated, and operated, buffer strips can provide reliable water quality benefits in conjunction with high aesthetic appeal.
- Flow characteristics and vegetation type and density can be closely controlled to maximize BMP effectiveness.
- Roadside shoulders act as effective buffer strips when slope and length meet criteria described in the subsequent sections.

8.3 Limitations

Buffer strips have some limitations, including the following:

- They cannot treat a very large drainage area.
- If designed improperly (e.g., if longitudinal slope is too great), they have very little pollutant removal capability.
- A thick vegetative cover is needed for trim to function properly.
- They are impractical in areas with steep topography, or wet or poorly drained soils.
- They are not effective and may even erode when flow volumes and/or velocities are high.
- Sufficient land may not be available for suitable buffer strip designs to be incorporated.
- Many local municipalities/councils in the United States and the United Kingdom may require curb and gutter systems in residential areas.
- They are impractical in areas with erosive soils or where a dense vegetative cover is difficult to maintain.
- Infiltration through the filter strip may carry pollutants into local groundwater.
- Filter strips require sheet flow. Piping surface water to the strip is not recommended as this may cause erosion and minimize filtration opportunities.

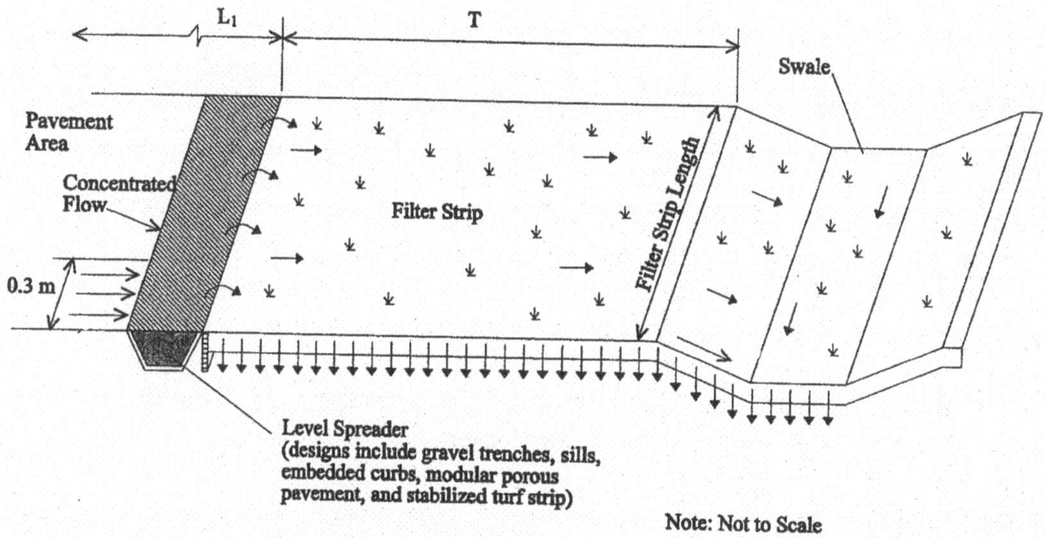

Figure 8-1. Engineered Filter Strip Schematic.

Figure 8-2. Filter Strip, Hopwood Services, United Kingdom.

8.4 Siting Criteria

The use of natural or engineered biofilters is limited to gently sloping areas where the vegetative cover is robust and diffuse, and where shallow flow characteristics are possible. The practical water quality benefits of each practice can be effectively eliminated with the occurrence of significant erosion or when flow concentration occurs across the vegetated surface. Slopes should not exceed 10% or be less than 1%. The vegetative surface should extend across the full width of the area being drained. The upstream boundary of the filter should be located contiguous to the developed area. Use of a level spreading device (e.g., vegetated berm, sawtooth concrete border, rock trench) to facilitate overland sheet flow is not normally recommended because of maintenance considerations and the potential for standing water.

Filter strips are applicable in most regions, but are restricted in some situations because they consume a large amount of space relative to other practices. Filter strips are best suited to treating runoff from roads and highways, roof downspouts, very small parking lots, and pervious surfaces. They are also ideal components of the outer zone of a stream buffer, or as pretreatment to a structural practice. In arid areas, however, the cost of irrigating the grass on the practice will most likely outweigh its water quality benefits, although aesthetic considerations may be sufficient to overcome this constraint. Ultra-urban areas are densely developed urban areas in which little pervious surface exists. Filter strips are impractical in these areas because of space constraints. Filter strips also are generally a poor retrofit option because they consume a relatively large amount of space and cannot treat large drainage areas.

Some cold water species, such as trout, are sensitive to changes in temperature. Although some treatment practices, such as wet ponds, can warm stormwater substantially, filter strips do not warm pond water on the surface for long periods of time and are not expected to increase stormwater temperatures. Therefore, these practices are good for protection of cold-water streams.

8.5 Design Guidelines

Filter strips appear to be a minimal design practice because they are basically no more than a grassed slope.

A major question that remains unresolved is how large the drainage area to a strip can be. Research has conclusively demonstrated that these are effective on roadside shoulders, where the contributing area is about twice the buffer area. They have also been installed on the perimeter of large parking lots where they performed fairly effectively; however, much lower slopes may be needed to provide adequate water quality treatment.

Table 8-1. Design Guidelines.

Design Criteria	Current Guidance/Recommendations
Planting	♦ The filter area should be densely vegetated with a mix of erosion-resistant plant species that effectively bind the soil. Native or adapted grasses, shrubs, and trees are preferred because they generally require less fertilizer and are more drought resistant than exotic plants.
Slope	♦ In general the slope of the strip should not exceed 10%. The gentler the slope, the greater its effectiveness for pollutant removal. ♦ Both the top and toe of the slope should be as flat as possible to encourage sheet flow and prevent erosion.
Filter Strip Length	♦ The strip should be at least 5–7 m (16–23 ft) long to provide water quality treatment.
Flow Velocities	♦ Runoff flow velocities should not exceed about 0.3 m/s (1 ft/s) across the vegetated surface to encourage settlement during regular storms. ♦ Flow velocities for larger floods should be below 1.5 m/s (5 ft/sec) to prevent erosion.
Inlet Infrastructure	♦ The top of the strip should be installed 50–125 mm (2–5 in.) below the adjacent pavement, so that vegetation and sediment accumulation at the edge of the strip does not prevent runoff from entering.
Subsoil	♦ Topsoil should be lightly compacted to a depth of 100 mm (4 in.).
Water Table	♦ Geotechnical tests must be performed to determine the location of the water table. If the water table is within 0.6 m (2 ft) of the planned swale bottom, a dry swale is not feasible. ♦ If there is a risk of a swale conveying potential polluted water, and the groundwater requires protection, an impermeable liner should be incorporated in the design (CIRIA, 2000a,b)
Groundwater Levels	♦ Filter strips should be separated from the groundwater by between 0.6–1.2 m (2–4 ft) to prevent contamination and to ensure that the filter strip does not remain wet between storms.

For engineered vegetative strips, the facility surface should be graded flat prior to placement of vegetation. Initial establishment of vegetation requires attentive care including appropriate watering, fertilization and prevention of excessive flow across the facility until vegetation completely covers the area and is well established. Use of a permanent irrigation system may help provide maximal water quality performance.

In cold climates, filter strips provide a convenient area for snow storage and treatment. If used for this purpose, vegetation in the filter strip should be salt-tolerant, (e.g., creeping bentgrass), and a maintenance schedule should include the removal of sand built up at the bottom of the slope. In arid or semiarid climates, designers should specify drought-tolerant grasses (e.g., buffalo grass) to minimize irrigation requirements.

8.6 Construction Guidelines

- If possible, divert runoff (other than necessary irrigation) during the period of vegetation establishment. Where runoff diversion is not possible, cover graded and seeded areas with suitable erosion-control materials.
- The filter strip should be carefully prepared to ensure that it is structurally sound, level, and true and will perform its required hydraulic function. The strip should not be compacted during construction.

8.7 Performance

Vegetated buffer strips tend to provide somewhat better treatment of stormwater runoff than swales and are less prone to channelization or erosion. Table 8-2 documents the pollutant removal observed in a recent study by Caltrans (2002) of three sites in southern California. The column labeled "Significance" represents the probability that the mean influent and effluent EMCs are not significantly different based on an ANOVA.

The removal of sediment and dissolved metals was comparable to that observed in much more complex controls. Reduction in nitrogen was not significant and all of the sites exported phosphorus for the entire study period. This may have been the result of using salt grass, a warm weather species that is dormant during the wet season.

Another Caltrans study (unpublished) of vegetated highway shoulders as buffer strips also found substantial reductions, often within a very short distance of the edge of pavement. Figure 8-3 presents the concentrations of TSS in highway runoff after traveling various distances (shown in meters) through a vegetated filter strip with a slope of about 10%. One can see that the TSS reaches an irreducible minimum concentration of about 20 mg/L within 5 m (16 ft) of the pavement edge.

Filter strips also exhibit good removal of litter and other floatables because the water depth in these systems is well below the vegetation height and consequently these materials are not easily transported through them. Unfortunately little attenuation of peak runoff rates and volumes (particularly for larger events) is normally observed, depending on the soil properties. Therefore, it may be prudent to follow the strips with another practice that can reduce flooding and channel erosion downstream.

Table 8-2. Pollutant Reduction in a Vegetated Buffer Strip.

Constituent	Mean EMC Influent mg/L	Mean EMC Effluent mg/L	Removal %	Significance P
TSS	119	31	74	<.000
NO_3-N	0.67	0.58	13	.367
TKN-N	2.50	2.10	16	.542
Total N	3.22	2.83	12	-
Dissolved P	0.15	0.46	-206	.047
Total P	0.42	0.62	-52	.035
Total Cu	0.058	0.009	84	<.000
Total Pb	0.046	0.006	88	<.000
Total Zn	0.245	0.055	78	<.000
Dissolved Cu	0.029	0.007	77	.004
Dissolved Pb	0.004	0.002	66	.006
Dissolved Zn	0.099	0.035	65	<.000

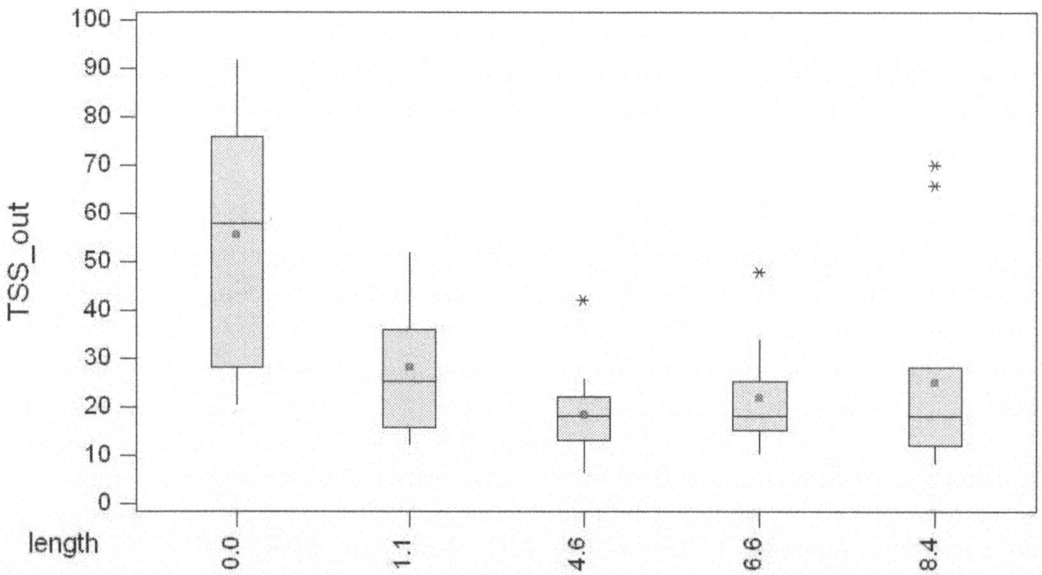

Figure 8-3. TSS Concentration in Buffer Strip Discharge (length in meters). Asterisks represent data outliers.

8.8 Maintenance

Filter strips require mainly vegetation management; therefore little special training is needed for maintenance crews. Typical maintenance activities and frequencies are listed in the following table.

Table 8-3. Typical Maintenance Activities for Filter Strips.

Activity	Schedule
♦ Inspect strips at least twice annually for erosion or damage to vegetation, preferably at the end of the wet season to schedule summer maintenance and before major fall runoff to be sure the strip is ready for winter. However, additional inspection after periods of heavy runoff is most desirable to check for rills, gullies, or water logging.	Biannually
♦ The strip should be checked regularly for debris, litter, and areas of sediment accumulation. ♦ Trash tends to accumulate in strip areas, particularly along highways. The need for litter removal should be determined through periodic inspection, but litter should always be removed prior to mowing.	As needed
♦ Recent research on biofiltration swales, but likely applicable to strips (Colwell et al., 2000) indicates that grass height and mowing frequency have little impact on pollutant removal; consequently, mowing may only be necessary once or twice a year for safety and aesthetics or to suppress weeds and woody vegetation.	Annually
♦ Sweeping and erosion control of the upstream catchment should be undertaken to reduce opportunities for silt to enter the system.	Monthly

8.9 Design Life

It has been suggested that with adequate maintenance, a design life of 20–50 years might be appropriate.

8.10 Costs

8.10.1 Construction Costs

Limited data are available on the actual construction costs of filter strips. One rough estimate can be the cost of seed or sod, which is approximately $3.20/m^2 ($0.30/ft^2) for seed or $7.50/m^2 ($0.70/ft^2) for sod. This amounts to between $13,000 and $30,000 per acre of filter strip. This cost is relatively high compared with other treatment practices. However, the grassed area used as a filter strip may have been seeded or sodded even if it were not used for treatment. In these cases, the only additional costs are the design.

The true cost of filter strips is the cost of the land area they occupy, which is greater than that of any other treatment practice. In some situations this land is available as unused space beyond backyards or adjacent to roadsides, but filter strips are cost prohibitive where land prices are high and the land could be used for other purposes.

8.10.2 Maintenance Costs

Maintenance of vegetated buffer strips consists mainly of vegetation management (mowing, irrigation if needed, weeding) and litter removal. Consequently the costs are quite variable depending on the frequency of these tasks and the local labor rate.

Typical maintenance costs are about $350/ac/yr (adapted from SWRPC, 1991). This cost is relatively moderate and might overlap with regular landscape maintenance costs.

8.11 Research Needs

8.11.1 Design Guidelines

- Size of drainage area.

8.11.2 Performance

- Effects of mowing on pollutant removal capability of grassed filter strips;
- Effects of different grass types on pollutant removal capability; and
- Whether removal rates change with age.

8.11.3 Maintenance

- Optimum frequency to retain optimum grass lengths.

8.11.4 Costs

- Construction out-turn (as-built) costs required.

8.12 Additional Resources

Center for Watershed Protection (CWP), 1997, *Stormwater BMP Design Supplement for Cold Climates*. Prepared for U.S. Environmental Protection Agency Office of Wetlands, Oceans and Watersheds. Washington, D.C.

Desbonette, A., Pogue, P., Lee, V., and Wolff, N., 1994, *Vegetated Buffers in the Coastal Zone: A Summary Review and Bibliography*. Coastal Resources Center. University of Rhode Island, Kingston, RI.

Magette, W., Brinsfield, R., Palmer, R., and Wood, J., 1989, Nutrient and Sediment Removal by Vegetated Filter Strips. *Transactions of the American Society of Agricultural Engineers* 32(2): 663–667.

CHAPTER 9.0

SAND AND ORGANIC FILTERS

9.1 Description

Sand filters are usually two-chambered stormwater practices; the first is a settling chamber, and the second is a filter bed filled with sand or another filtering media. As stormwater flows into the first chamber, large particles settle out, and then finer particles and other pollutants are removed as stormwater flows through the filtering medium.

The use of slow-sand filters for the treatment of stormwater runoff is a fairly recent innovation. The filtration of water through sand as a means of improving its quality was first performed in 1829 in London to treat Thames River water. This first filter was the predecessor of the slow-sand type later developed in England and used extensively at the beginning of the 20th century in the United States for water and wastewater treatment. Over the intervening years, the use of slow-sand filters for the treatment of water and wastewater declined as improved rapid filtering and treatment technologies were developed (Young et al., 1996).

Since their original inception in Austin, Texas, hundreds of intermittent sand filters have been implemented to treat stormwater runoff. There have been numerous alterations or variations in the original design as engineers in other jurisdictions have improved and adapted the technology to meet their specific requirements. Major types include the Austin Sand Filter (Figure 9-1), the District of Columbia Underground Sand Filter, the Alexandria Dry Vault Sand Filter, the Delaware Sand Filter (Figure 9-2), and Multichamber Treatment Train (Figure 9-3). The primary differences among these designs are location (i.e., above or below ground), the drainage area served, their filter surface areas, their land requirements, and the quantity of runoff they treat. Modifications to the traditional surface sand filter were made primarily to fit sand filters into more challenging design sites (e.g., underground and perimeter filters) or to improve pollutant removal (e.g., organic media filter).

The City of Austin recognizes two design alternatives: full and partial sedimentation. In the full sedimentation design the sedimentation basin is sized to contain the entire water quality volume, which is then released over 24 hours to the filter basin. The partial sedimentation design consists of a single basin, sized to contain the water quality volume, and is underlain by the filter bed. Because the amount of pretreatment in the partial sedimentation design is low, Austin specifies a lower permeability for the sand, which results in a larger filter area as a safety factor.

9.2 Advantages

- Appropriate for areas where space is limited;
- Applicable in arid climates where wet basins and constructed wetlands are not appropriate; and
- High TSS removal efficiency.

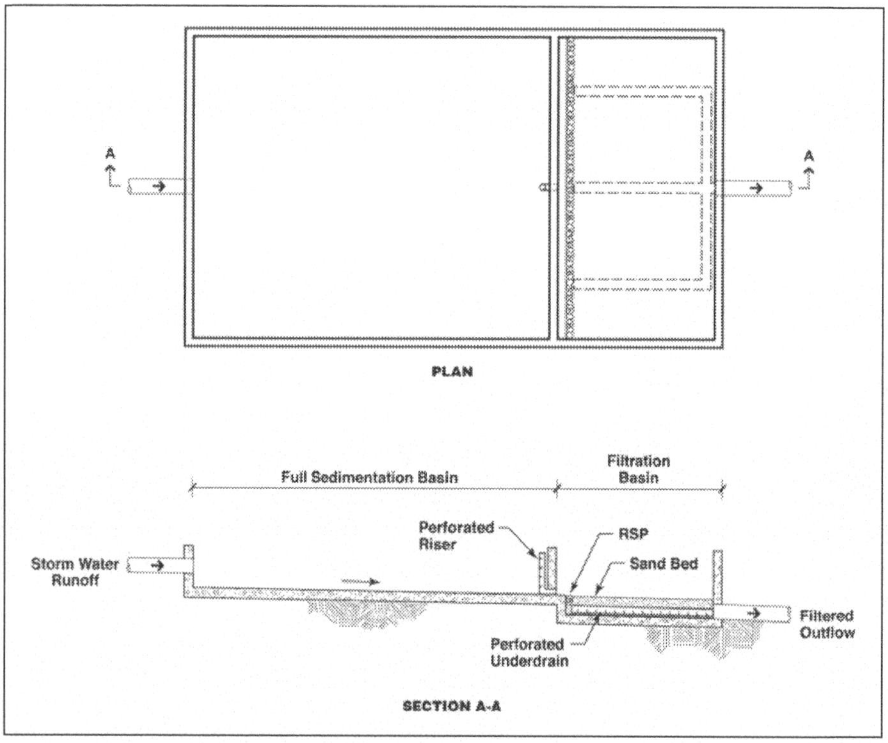

Figure 9-1. Schematic of Full Sedimentation Austin Sand Filter.

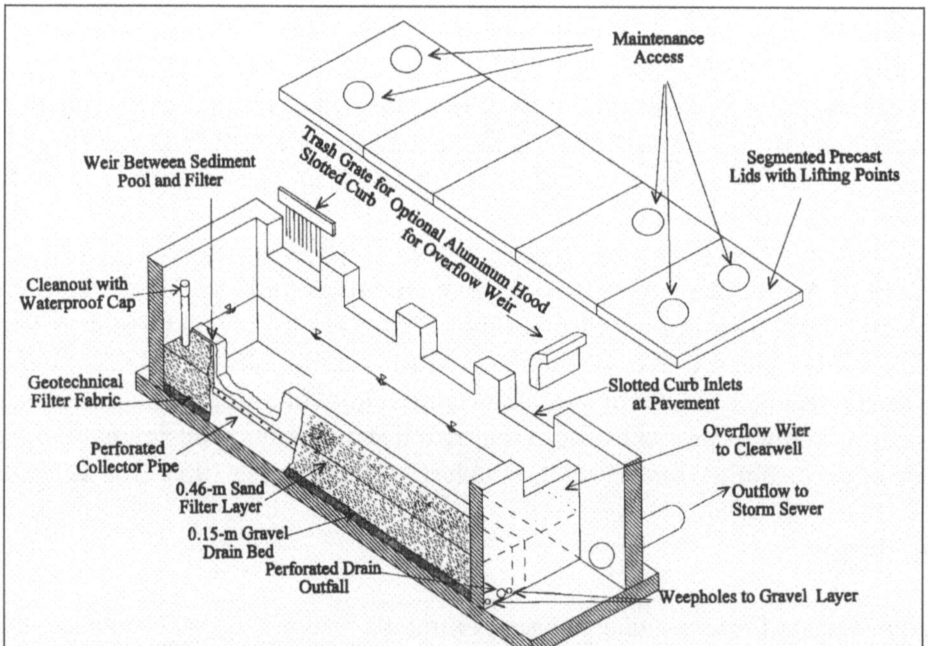

Figure 9-2. Schematic of Delaware Sand Filter (Young et al., 1996).

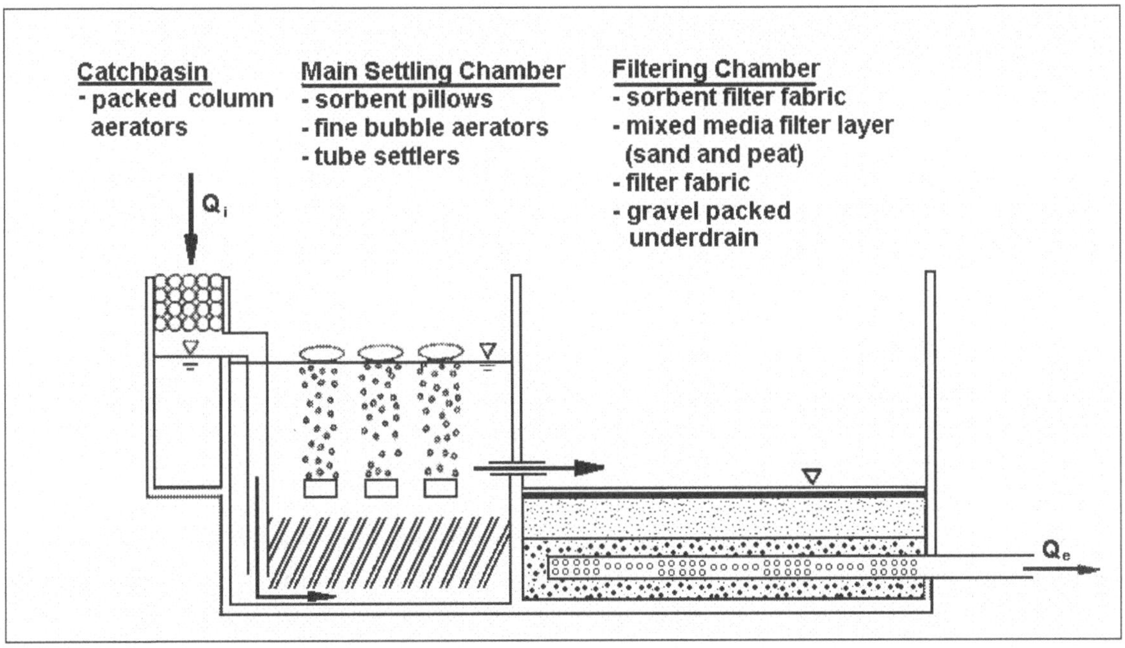

Figure 9-3. Schematic of Multichambered Treatment Train.

9.3 Limitations

- Require more maintenance that most other BMPs depending upon the sizing of the filter bed;
- Generally require more hydraulic head to operate properly (minimum 4 ft);
- High solids loads will cause the filter to clog;
- Work best for relatively small, impervious watersheds; and
- Filters in residential areas can present aesthetic and safety problems.

9.4 Siting Criteria

In general, sand filters are preferred over infiltration practices, such as infiltration trenches, when contamination of groundwater with conventional pollutants—BOD, suspended solids, and fecal coliform—is of concern. This usually occurs in areas where underlying soils alone cannot treat runoff adequately or where groundwater tables are high. In most cases, sand filters can be constructed with impermeable basin or chamber bottoms, which help to collect, treat, and release runoff to a storm drainage system or directly to surface water with no contact between contaminated runoff and groundwater. In regions where evaporation exceeds rainfall and a wet pond would be unlikely to maintain the required permanent pool, the Austin sand filtration system can be used.

The selection of a sand filter design depends largely on the drainage area's characteristics. For example, the Washington, D.C. and Delaware sand filter systems are well suited for highly impervious areas where land available for structural controls is limited, since both are installed underground. They are often used to treat runoff from parking lots, driveways, loading docks, service stations, garages, airport runways/taxiways, and storage yards. The Austin sand filtration system is more suited for large drainage areas that have both impervious and pervious surfaces. This system is located at grade and is often used at transportation facilities, in large parking areas, and in commercial developments.

It is challenging to use most sand filters in very flat terrain because they require a significant amount of hydraulic head (1.2 m [about 4 ft]), to allow flow through the system. One exception is the perimeter sand filter, which can be applied with as little as 0.6 m (2 ft) of head.

Sand filters are best applied on relatively small sites (10 ha [up to 25 ac]) for surface sand filters and closer to 0.8 ha (2 ac) for perimeter or underground filters)]. Filters have been used on larger drainage areas, of up to 40 ha (100 ac), but these systems can clog when they treat larger drainage areas unless adequate measures are provided to prevent clogging, such as a larger sedimentation chamber or more intensive regular maintenance.

When sand filters are designed as a stand-alone practice, they can be used on almost any soil because they can be designed so that stormwater never infiltrates into the soil or interacts with the ground water. Alternatively, sand filters can be designed as pretreatment for an infiltration practice, where soils do play a role.

9.5 Design Guidelines

Pretreatment is a critical component of any stormwater management practice. In sand filters, pretreatment is achieved in the sedimentation chamber that precedes the filter bed. In this chamber, the coarsest particles settle out and thus do not reach the filter bed. Pretreatment reduces the maintenance burden of sand filters by reducing the potential of these sediments to clog the filter. Designers should provide at least 25% of the water quality volume in a dry or wet sedimentation chamber as pretreatment to the filter system. The water quality volume is the amount of runoff that will be treated for pollutant removal in the sedimentation chambers and filter bed.

Treatment design features help enhance the ability of a stormwater management practice to remove pollutants. In filtering systems, designers should provide at least 75% of the water quality volume in the sedimentation chambers and filter bed. In sand filters, designers should select a medium sand as the filtering medium. A fine aggregate (ASTM C-33) that is intended for use in concrete is commonly specified.

The filter bed should be sized using D'Arcy's Law, which relates the velocity of fluids to the hydraulic head and the coefficient of permeability of a medium. The resulting equation, as derived by the City of Austin, Texas, (1996b), is

$$AF = WTV\ d/[k\ t\ (h+d)]$$

Where: WTV = Water treatment volume, ft^3

AF = area of the filter bed (ft^2);

d = depth of the filter bed (ft; usually about 1.5 ft, depending on the design);

k = coefficient of permeability of the filtering medium (ft/day);

t = time for the water quality volume to filter through the system (days; usually assumed to be 1.67 days); and
h = average water height above the sand bed (ft; assumed to be one-half of the maximum head);
Typical values for k, as assembled by CWP (1996), are shown in Table 9-1.

The value for permeability of sand shown in Table 9-1 is extremely conservative but is widely used since it is incorporated in the design guidelines of the City of Austin. When the sand is initially installed, the permeability is so high that generally only a portion of the filter area is required to infiltrate the entire volume, especially in a full-sedimentation Austin design where the capture volume is released to the filter basin over 24 hours.

Table 9-1. Coefficient of Permeability Values for Stormwater Filtering Practices (CWP, 1996).

Filter Medium	Coefficient of Permeability (cm/day)	Coefficient of Permeability (ft/day)
Sand	1.1	3.5
Peat/Sand	0.84	2.8
Compost	2.6	8.7

Typically, filtering practices are designed as off-line systems, meaning that during larger storms all runoff greater than the water quality volume is bypassed untreated using a flow splitter, which is a structure that directs larger flows to the storm drain system or to a stabilized channel. One exception is the perimeter filter; in this design, all flows enter the system, but larger flows overflow to an outlet chamber and are not treated.

All filtering practices, with the exception of exfiltration devices (see "Design Variations," below) are designed with an under drain below the filtering bed. An under drain is a perforated pipe system in a gravel bed, installed on the bottom of filtering practices and used to collect and remove filtered runoff.

9.5.1 Design Variations

As mentioned earlier in this fact sheet, there are five basic stormwater filter designs — surface sand filter, underground filter, perimeter filter (also known as the "Delaware" filter), organic media filter, and Multichamber Treatment Train. Other design variations can incorporate design features to recharge groundwater or to meet the design challenges of cold or arid climates.

9.5.2 Surface Sand Filter

The surface sand filter is the original sand filter design. In this practice both the filter bed and the sediment chamber are aboveground. The surface sand filter is designed as an off-line practice, where only the water quality volume is directed to the filter. The surface sand filter is the least expensive filter option and has been the most widely used.

9.5.3 Underground Sand Filter

The underground sand filter is a modification of the surface sand filter, where all of the filter components are underground. Like the surface sand filter, this practice is an off-line system that receives only the smaller water quality events. Underground sand filters are expensive to construct but consume very little space. They are well suited to highly urbanized areas.

9.5.4 Perimeter Sand Filter

The perimeter sand filter also includes the basic design elements of a sediment chamber and a filter bed. In this design, however, flow enters the system through grates, usually at the edge of a parking lot. The perimeter sand filter is the only filtering option that is on-line, with all flows entering the system but larger events bypassing treatment by entering an overflow chamber. One major advantage to the perimeter sand filter design is that it requires little hydraulic head and thus is a good option in areas of low relief.

9.5.5 Organic Media Filter

Organic media filters are essentially the same as surface filters, with the sand medium replaced with or supplemented by another medium. Two examples are the peat/sand filter (Galli, 1990b) and the compost filter system (CSF, 1996). The assumption is that these systems will have enhanced pollutant removal for many compounds because of the increased cation exchange capacity achieved by increasing the organic matter in the filter media.

9.5.6 Multichamber Treatment Train

The multichamber treatment train (Robertson et al., 1995) is essentially a "deluxe sand filter." This underground system consists of three chambers. Stormwater enters the first chamber, where screening occurs, trapping large sediments and releasing highly volatile materials. The second chamber provides settling of fine sediments and further removal of volatile compounds and also floatable hydrocarbons through the use of fine bubble diffusers and sorbent pads. The final chamber provides filtration by using a sand and peat mixed medium for reduction of the remaining pollutants. The top of the filter is covered by a filter fabric that evenly distributes the water volume and prevents channelization. Although this practice can achieve very high pollutant removal rates, it may be prohibitively expensive in many areas and has been implemented only on an experimental basis.

9.5.7 Exfiltration/Partial Exfiltration

In these designs, all or part of the under drain system is replaced with an open bottom that allows infiltration to the ground water. When the under drain is present, it is used as an overflow device in case the filter becomes clogged. These designs are best applied in the same soils where infiltration practices are used.

In cold climates, filters can be used, but surface or perimeter filters will not be effective during the winter months, and unintended consequences might result from a frozen filter bed. Using alternative conveyance measures such as a weir system between the sediment chamber and filter bed may avoid freezing associated with the traditional standpipe. Where possible, the filter bed should be below the frost line. Some filters, such as the peat/sand filter, should be shut down during the winter. These media will become completely impervious during freezing conditions. Using a larger underdrain system to encourage rapid draining during the winter months may prevent freezing of the filter bed.

9.6 Performance

The pollutant removal performance of sand filters and other stormwater BMPs is generally characterized by the percent reduction in the influent load. This method implies a relationship between influent and effluent concentrations. For instance, it would be expected that a device that is reported to achieve a 75% reduction would have an effluent concentration equal to 25% of the influent concentrations. Recent work in California on Austin sand filters indicate that this model for characterizing performance is inadequate. Figure 9-4 presents a graph relating influent and effluent TSS concentrations for the Austin full sedimentation design.

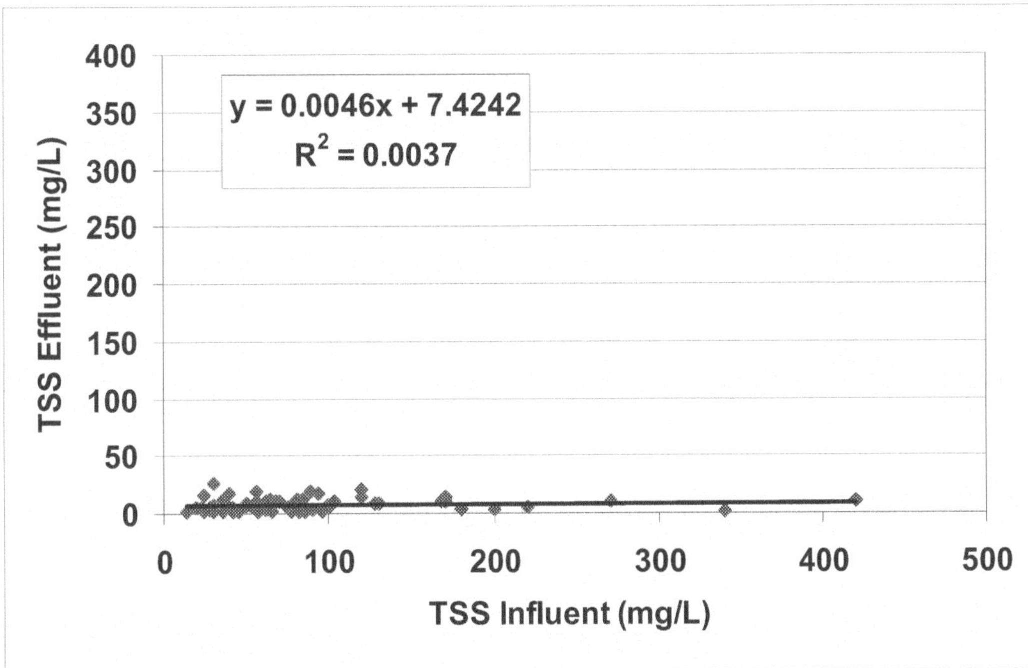

Figure 9-4. Comparison of Influent and Effluent TSS Concentrations.

It is clearly apparent that the effluent concentration is relative constant and independent of influent concentration. Consequently, the performance is more accurately characterized by the effluent concentration, which is about 7.5 mg/L. Constant effluent concentrations also are observed for all other particle-related constituents such as particulate metals (total-dissolved) and particulate phosphorus.

The small uncertainty in the estimate of the mean effluent concentration highlights the very consistent effluent quality for TSS produced by sand filters. In addition, it demonstrates that a calculated percent reduction for TSS and other constituents with similar behavior for Austin sand filters is a secondary characteristic of the device and depends primarily on the specific influent concentrations observed. The distinction between a constant effluent quality and a percent reduction is extremely important to recognize, if the results are to be used to estimate effluent quality from sand filters installed at other sites with different influent concentrations or for estimating compliance with water quality standards for storms with high concentrations of particulate constituents.

If the conventionally derived removal efficiency (90%) were used to estimate the TSS concentrations in the treated runoff from storms with high influent concentrations, the estimated effluent concentration would be too high. For instance, the storm with the highest observed influent concentration (420 mg/L) would be expected to have a concentration in the treated runoff of 42 mg/L, rather than the 10 mg/L that was measured. In fact, the TSS effluent concentrations for all events with influent concentrations greater than 200 mg/L were 10 mg/L or less.

The stable effluent concentration of a sand filter under very different influent TSS concentrations implies something about the properties of the influent particle size distribution. If one assumes that only the smallest size fraction can pass through the filter, then the similarity in effluent concentrations suggests that there is little difference in the total mass of the smallest sized particles even when the total TSS concentration varies greatly. Further, the difference in TSS concentration must then be caused by changes in the relative amount of the larger size fractions. Further research is necessary to determine the range of particle size that is effectively removed in the filter and the portion of the size fraction of suspended solids that it represents in urban stormwater.

Sand filters are effective stormwater management practices for pollutant removal. Conventional removal rates for all sand filters and organic filters are presented in Table 9-2. With the exception of nitrates, which are always exported from filtering systems because of the conversion of ammonia and organic nitrogen to nitrate, they perform relatively well at removing pollutants.

Table 9-2. Sand Filter Removal Efficiencies (%).

		Compost Filter System		**Multichamber Treatment Train**		
	Sand Filter (Schueler, 1997a)	**Stewart, 1992**	**Leif, 1999**	**Pitt, Lilburn, and Burian, 1997**	**Pitt, 1996**	**Greb, Corsi, and Wasch-bush, 1998**
TSS	87	95	85	85	83	98
TP	51	41	4	80	—	84
TN	44	—	—	—	—	—
Nitrate	-13	-34	-95	—	14	—
Metals	34-80	61-88	44-75	65-90	91-100	83-89
Bacteria	55	—	—	—	—	—

From the few studies available, it is difficult to determine if organic filters necessarily have higher removal efficiencies than sand filters. The Multichamber Treatment Train may have high pollutant removal for some constituents, although an evaluation of these devices by the California Department of Transportation indicated no significant difference for most conventional pollutants.

9.7 Maintenance

Even though sand filters are generally thought of as one of the higher maintenance BMPs, in a recent California study an average of only about 49 hours a year was required for field activities. This was less maintenance than was required by extended detention basins serving comparable sized catchments. Most maintenance consists of routine removal of trash and debris, especially in Austin sand filters where the outlet riser from the sedimentation basin can become clogged.

Most data (e.g., Clark, 2000) indicate that hydraulic failure from clogging of the sand media occurs before pollutant breakthrough. Typically, only the very top of the sand becomes clogged while the rest remains in relative pristine condition as shown in Figure 9-5. The rate of clogging has been related to the TSS loading on the filter bed (Urbonas, 1999); however, the data are quite variable. Empirical observation of sites treating urban and highway runoff indicates that clogging of the filter occurs after 2–10 years of service. Presumably, this is related to differences in the type and amount of sediment in the catchment areas of the various installations. Once clogging occurs the top 2–3 in. of filter media is removed, which restores much, but not all, of the lost permeability. This removal of the surface layer can occur several times before the entire filter bed must be replaced. The cost of the removal of the surface layer is not prohibitive, generally ranging between $2000 (U.S. EPA, 1999f) and $4000 (Caltrans, 2002), depending on the size of the filter.

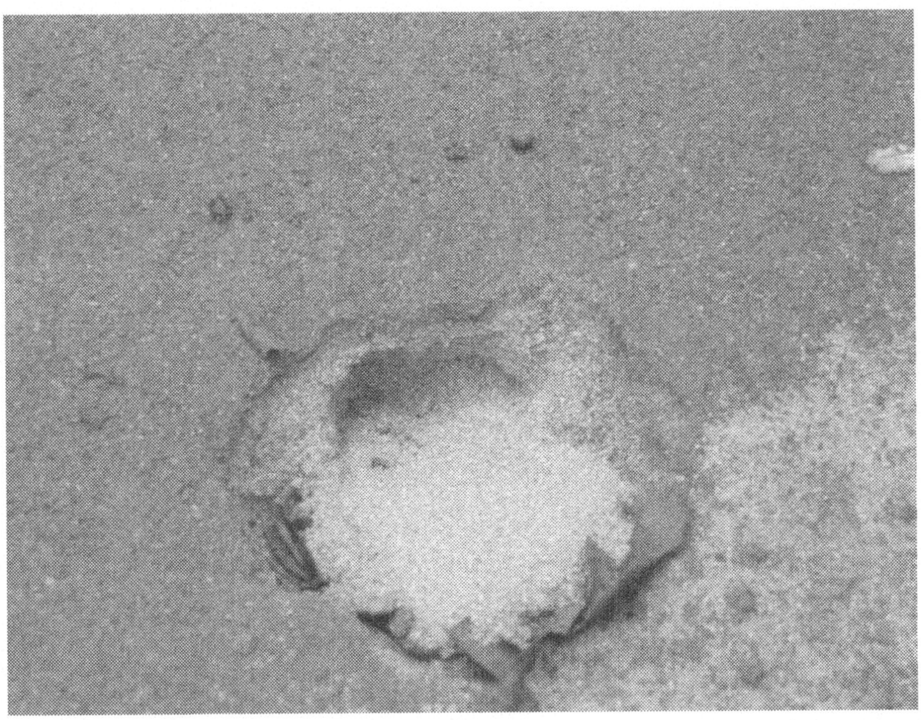

Figure 9-5. Formation of Clogging Crust on Filter Bed.

Recommended maintenance activities and frequencies include:

Table 9-3. Typical Maintenance Activities for Sedimentation Basin and Filter Bed.

Activity	Schedule
♦ Inspect for standing water, sediment, trash, and debris, and to identify potential problems.	Semiannually
♦ Remove accumulated trash and debris in the sedimentation basin, from the riser pipe, and the filter bed during routine inspections.	Semiannually
♦ Inspect the facility once during the wet season after a large rain event to determine whether the facility is draining completely within 72 hours.	Annually
♦ Remove top 50 mm (2 in.) of sand and dispose of sediment if facility drain time exceeds 72 hours. Restore media depth to 450 mm (18 in.) when overall media depth drops to 300 mm (12 in.).	As needed
♦ Remove accumulated sediment in the sedimentation basin every 10 years or when the sediment occupies 10% of the basin volume, whichever is less.	Every 10 years or less

9.8 Costs

9.8.1 Construction Costs

There are few consistent published data on the cost of sand filters, largely because, with the exception of Austin, Texas; Alexandria, Virginia; and Washington, D.C., they have not been widely used. Furthermore, filters have such varied designs that it is difficult to assign a cost to filters in general. A study by Brown and Schueler (1997) was unable to find a statistically valid relationship between the volume of water treated in a filter and the cost of the practice. The EPA Fact Sheet (1999f) indicates a cost for an Austin sand filter at $18,500 (1997 dollars) for a 0.4 ha (1 ac) drainage area. However, the same design implemented at a 1.1 ha (0.45 ac) site by the California Department of Transportation, cost $240,000. Consequently, there is a tremendous uncertainty about what the average construction cost might be.

It is important to note that although underground and perimeter sand filters can be more expensive than surface sand filters, they consume no surface space, making them a relatively cost-effective practice in ultra-urban areas where land is at a premium.

Given the number of facilities installed in the areas that promote their use it should be possible to develop fairly accurate construction cost numbers through a more comprehensive survey of municipalities and developers that have implemented these filters.

9.8.2 Maintenance Costs

Annual costs for maintaining sand filter systems average about 5% of the initial construction cost (Schueler, 1992). Media is replaced as needed, with the frequency correlated with the solids loading on the filter bed. Currently the sand is being replaced in the D.C. filter systems about every two years, while an Austin design might last 3 to 10 years, depending on the watershed characteristics. The cost to replace the gravel layer, filter fabric, and top portion of the sand for D.C. sand filters is approximately $1700 (1997 dollars).

Caltrans estimated future maintenance costs for the Austin design, assuming a device sized to treat runoff from approximately 1.6 ha (4 ac). These estimates are presented in Table 9-4 and assume a fully burdened hourly rate of $44 for labor. This estimate is somewhat uncertain because complete replacement of the filter bed was not required during the period that maintenance costs were recorded.

Table 9-4. Expected Annual Maintenance Costs for an Austin Sand Filter.

Activity	Labor Hours	Equipment and Materials ($)	Cost
Inspections	4	0	176
Maintenance	36	125	1,706
Vector Control	0	0	0
Administration	3	0	132
Direct Costs	-	888	888
Total	43	$1,013	**$2,902**

9.9 Research Needs

Sand filters have been widely used and monitored in a few areas of the United States, so their siting criteria, design guidelines, and performance for most common constituents is well understood. One area related to performance that could be investigated is the use of media other than sand to improve pollutant removal for dissolved constituents such as nutrients and metals.

There is little documentation of actual maintenance activities and their associated costs. Additional work in this area is recommended.

Data reported on construction costs for these devices is highly variable, so some additional effort to identify the reasons for the wide ranges of costs reported is warranted.

9.10 Additional Resources

AquaLogic Stormwater Filtration System Engineering Manual, 2000, AquaLogic Stormwater Abatement Filter Systems. San Antonio, TX.

Barton Springs/Edwards Aquifer Conservation District, 1996, *Final Report: Enhanced Roadway Runoff Best Management Practices*. City of Austin, Drainage Utility, LCRA, TDOT. Austin, TX.

Bell, W., L. Stokes, L.J. Gavan, and T.N. Nguyen, 1995, *Assessment of the Pollutant Removal Efficiencies of Delaware Sand Filter BMPs*. Final Report. Department of Transportation and Environmental Services. Alexandria, VA. Also in Performance of Delaware Sand Filter Assessed. *Watershed Protection Techniques*.

Center for Watershed Protection (CWP), 1997, *Stormwater BMP Design Supplement for Cold Climates*. Prepared for U.S. EPA Office of Wetlands, Oceans and Watersheds, Washington, D.C., by the Center for Watershed Protection, Ellicott City, MD.

City of Alexandria, Virginia, no date, Unconventional BMP Design Criteria. In *Alexandria Supplement to the Northern Virginia BMP Handbook*. Alexandria, VA.

Curran, T., 1996, Peat Sand Efficiency Calculations for McGregor Park. Unpublished data. Lower Colorado River Authority. Austin, TX.

Harper, H., and Herr, J., 1993, *Treatment Efficiency of Detention with Filtration Systems*. Environmental Research and Design, Inc. Final Report Submitted to Florida Department of Environmental Regulation. Orlando, FL.

Horner, R.R., and Horner, C.R., 1999, Performance of a Perimeter ("Delaware") Sand Filter in Treating Stormwater Runoff from a Barge Loading Terminal. *Proc. of the Comprehensive Stormwater and Aquatic Ecosystem Management Conf.*, Auckland, N.Z., Feb. 1999, pp. 183-192.

Horner, R.R., and Horner, C.R., 1995, *Design, Construction and Evaluation of a Sand Filter Stormwater Treatment System*. Part II. *Performance Monitoring*. Report to Alaska Marine Lines, Seattle, WA. 38 p. Also in Performance of Delaware Sand Filter Assessed. *Watershed Protection Techniques*. Center for Watershed Protection. Fall 1995. Vol. 2(1): 291–293.

Keblin, M.V., Barrett, M.E., Malina, J.F., Jr., Charbeneau, R.J., 1998, *The Effectiveness of Permanent Highway Runoff Controls: Sedimentation/Filtration Systems*, Research Report 2954-1, Center for Transportation Research, University of Texas at Austin.

King County, Washington, Department of Natural Resources, 2000, *King County Surface Water Design Manual.* [http://splash.metrokc.gov/wlr/dss/manual.htm]. Last updated March 6, 2000. Accessed January 5, 2001.

Schueler, T., 1994, Developments in sand filter technology to improve stormwater runoff quality. *Watershed Protection Techniques* 1(2):47–54.

Shaver, E., and Baldwin, R., 1991, *Sand Filter Design for Water Quality Treatment*. Delaware Department of Natural Resources and Environmental Control, Dover, DE.

Washington State Department of Ecology (DOE), 1992, *Stormwater Management Manual for the Puget Sound Basin*, Washington State Department of Ecology, Olympia, WA.

CHAPTER 10.0

WATER QUALITY INLETS/ OIL WATER SEPARATORS

10.1 Description

Water quality inlets (WQIs), also commonly called trapping catch basins, oil/grit separators, or oil/water separators, consist of one or more chambers that promote sedimentation of coarse materials and separation of free oil (as opposed to emulsified or dissolved oil) from stormwater. Some WQIs also contain screens to help retain larger or floating debris, and many of the newer designs include a coalescing unit that promote oil/water separation. A typical WQI, as shown in Figure 10-1, consists of a sedimentation chamber, an oil separation chamber, and a discharge chamber. In the United Kingdom a simple version of a WQI, termed a gullypot, is often incorporated in urban and highway drainage systems, but has so little volume that substantial removal of pollutants is unlikely.

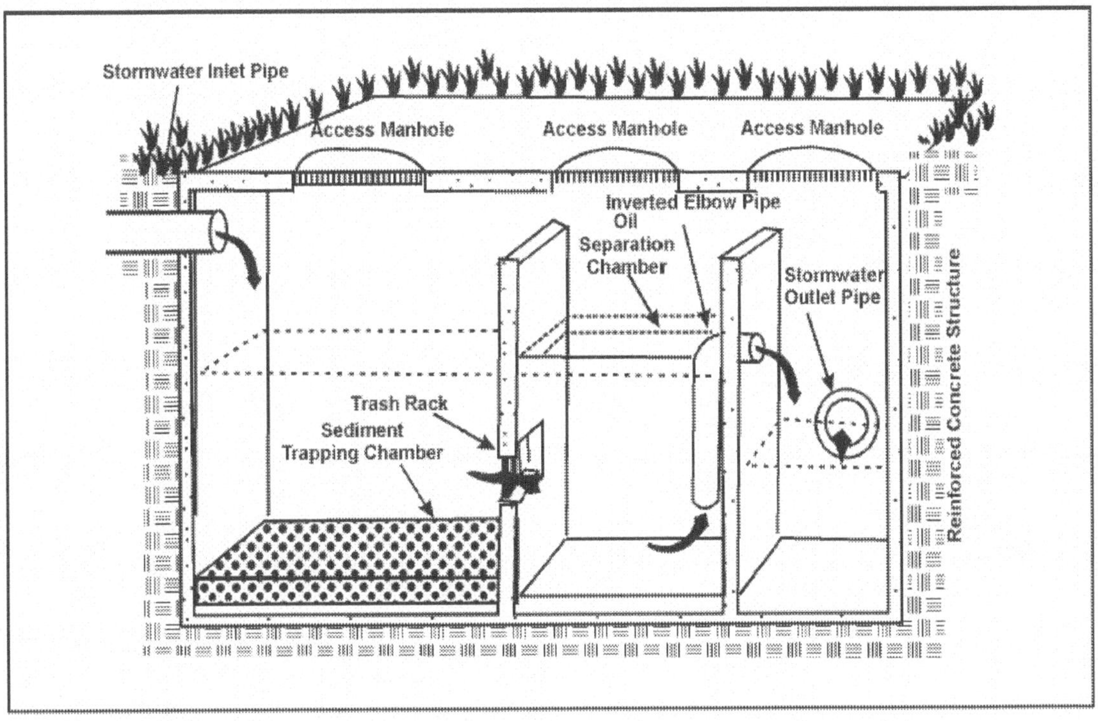

Figure 10-1. Schematic of Water Quality Inlet (Berg, 1991).

10.2 Advantages

- Water quality inlets and oil/water separators are effective for capturing hydrocarbon spills.

10.3 Limitations

- WQIs generally provide limited hydraulic and residuals storage. Due to the limited storage, WQIs provide only marginal sediment removal and stormwater treatment.
- Standing water in the devices can provide a breeding ground for mosquitoes.
- Even ideally designed configurations cannot remove pollutants as well as other structural stormwater management practices, such as wet ponds, sand filters, and wet ponds.
- Unless frequently maintained, these devices can become a source of pollutants through resuspension.
- Typical concentrations of oil and grease in stormwater runoff may be below the levels required for successful operation of these devices.

10.4 Design and Sizing Guidelines

- Water quality inlets are most effective for spill control and should be sized accordingly.

10.5 Performance

WQIs are most effective for collecting oil from spills, but also provide some removal of coarse sediment. A 1990 report by the American Petroleum Institute found that the efficiency of oil and water separation in a WQI is inversely proportional to the ratio of the discharge rate to the unit's surface area. Because of their small capacity, the discharge rate of the WQI is typically very high and their detention time is very short. For example, a study by the Metropolitan Washington Council of Governments (1993) found that the average detention time in a WQI is less than half an hour. This short detention can result in minimal pollutant settling (API, 1990). However, the addition of coalescing units in many current WQI units may increase oil/water separation efficiency. Most coalescing units are designed to achieve a specific outlet concentration of oil and grease (for example, 10–15 mg/L oil and grease).

Grit and sediment are partially removed by gravity settling within the first two chambers. A WQI with a detention time of 1 hour may be expected to remove 20% to 40% of solids. Hydrocarbons associated with the accumulated sediments are also often removed from the runoff through this process. The WQI achieves slight, if any, removal of nutrients, metals, and organic pollutants other than free petroleum products (Schueler, 1992).

The 1993 MWCOG study discussed above found that on average less than 5 cm (2 in.) of sediments (mostly coarse-grained grit and organic matter) was trapped in WQIs. Hydrocarbon and total organic carbon (TOC) concentrations of the sediments averaged 8150 and 53,900 mg/kg, respectively. The mean hydrocarbon concentration in the WQI water column was 10 mg/L. The study also indicated that sediment accumulation did not increase over time, suggesting that the sediments become resuspended during storm events. The authors concluded that although the WQI effectively separates oil and grease from water, resuspension of the settled matter appears to limit removal efficiencies. Actual removal occurs only when the residuals are removed from the WQI (Schueler, 1992).

Pollutant removal in stormwater inlets can be somewhat improved using inserts, which are promoted as means of removing oil and grease, trash, debris, and sediment. Some inserts are designed to drop directly into existing catch basins, whereas others may require extensive retrofit construction.

10.6 Siting Criteria

Oil/water separation units are often utilized in specific industrial areas, such as airport aprons, equipment washdown areas, or vehicle storage areas. In these instances, runoff from the area of concern is usually be diverted directly into the unit, while all other runoff is sent to the storm drain downstream from the oil/water separator. Oil/water separation tanks are often fitted with diffusion baffles at the inlets to prevent turbulent flow from entering the unit and resuspending settled pollutants.

10.7 Design Guidelines

Prior to WQI design, the site should be evaluated to determine if another BMP would be more cost effective for removing the pollutants of concern. WQIs should be used when no other BMP is feasible. The WQI should be constructed near a storm drain network so that flow can be easily diverted to the WQI for treatment (NVPDC, 1992). Any construction activities within the drainage area should be completed before installation of the WQI, and the drainage area should be revegetated so that the sediment loading to the WQI is minimized.

WQIs are most effective for small drainage areas. Drainage areas of 0.4 ha (1 acre) or less are often recommended. WQIs are typically used in an off-line configuration (i.e., portions of runoff are diverted to the WQI), but they can be used as on-line units (i.e., receive all runoff). Generally, offline units are designed to handle the first 1.3 cm (0.5 in.) of runoff from the drainage areas. Upstream isolation/diversion structures can be used to divert the water to the offline structure (Schueler, 1992). On-line units receive higher flows that will likely cause increased turbulence and resuspension of settled material, thereby reducing WQI performance.

Oil/water separation tanks are often fitted with diffusion baffles at the inlets to prevent turbulent flow from entering the unit and resuspending settled pollutants. WQIs are available as premanufactured units, or they can be cast in place. Reinforced concrete should be used to construct below-grade WQIs. The WQIs should be watertight to prevent possible groundwater contamination.

10.8 Maintenance

Frequent maintenance is required to prevent resuspension of accumulated sediment and displacement of captured oil and grease during subsequent storm events. Typical activities include trash removal if a screen or other debris capturing device is used, and removal of sediment using a vacuum (gulley sucker or vactor truck) truck. Operators need to be properly trained in catch basin maintenance. Maintenance should include keeping a log of the amount of sediment collected and the date of removal. Some cities have incorporated the use of geographic information systems to track sediment collection and to optimize future catch basin cleaning efforts.

EPA (1999c) reports that one study concluded that catch basins can capture sediments up to approximately 60% of the sump volume. When sediment fills greater than 60% of their volume, catch basins reach steady state. Storm flows can then resuspend sediments trapped in the catch basin, and they sediments will bypass treatment. Frequent clean-out can retain the volume in the catch basin sump available for treatment of stormwater flows.

At a minimum, catch basins should be cleaned at least twice during the wet season. Two studies suggest that increasing the frequency of maintenance can improve the performance of catch basins, particularly in industrial or commercial areas. One study of 60 catch basins in Alameda County, California, found that increasing the maintenance frequency from once per year to twice per year could increase the total sediment removed by catch basins on an annual basis (Mineart and Singh, 1994). Annual sediment removed per inlet was 24 kg (54 lb) for annual cleaning, 32 kg (70 lb) for semiannual and quarterly cleaning, and 73 kg (160 lb) for monthly cleaning. For catch basins draining industrial uses, monthly cleaning increased total annual sediment collected to six times the amount collected by annual cleaning (82 kg vs 4 kg [180 lb vs 30 lb]). These results suggest that, at least for industrial uses, more frequent cleaning of catch basins may improve efficiency.

10.9 Costs

A typical precast catch basin costs between $2000 and $3000; however, oil/water separators can be much more expensive. The true pollutant removal cost associated with catch basins, however, is the long-term maintenance cost. A vacuum truck, the most common method of catch basin cleaning, costs between $125,000 and $150,000. This initial cost may be high for smaller Phase II communities. However, it may be possible to share a vacuum truck with another community. Typical trucks can store between 7.6 and 11 m^3 (10 and 15 yd^3) of material, which is enough storage for three to five catch basins. Assuming semiannual cleaning and that the truck could be filled and material disposed of twice in one day, one truck would be sufficient to clean between 750 and 1000 catch basins. Another maintenance cost is the staff time needed to operate the truck. Depending on the regulations within a community, disposal costs of the sediment captured in catch basins may be significant.

10.10 Additional Resources

Aronson, G., Watson, D., and Pisaro, W., 1983, *Evaluation of Catch Basin Performance for Urban Stormwater Pollution Control.* U.S. Environmental Protection Agency, Washington, D.C.

Lager, J., Smith, W., Finn, R., and Finnemore, E., 1977, *Urban Stormwater Management and Technology: Update and User's Guide.* Prepared for U.S. Environmental Protection Agency. EPA-600/8-77-014. 313 pp.

Pitt, R., and Bissonnette, P., 1984, *Bellevue Urban Runoff Program Summary Report*. U.S. Environmental Protection Agency, Water Planning Division, Washington, D.C.

Pitt, R., Lilburn, M., Nix, S., Durrans, S.R., Burian, S., Voorhees, J., and Martinson, J., 2000, *Guidance Manual for Integrated Wet Weather Flow (WWF) Collection and Treatment Systems for Newly Urbanized Areas (New WWF Systems)*. U.S. Environmental Protection Agency, Office of Research and Development, Cincinnati, OH.

U.S. Environmental Protection Agency (U.S. EPA), 1999, *Stormwater Technology Fact Sheet: Water Quality Inlets,* EPA 832-F-99-029, Office of Water, Washington D.C.

CHAPTER 11.0

BIORETENTION

11.1 Description

Bioretention functions as a soil and plant-based filtration device that removes pollutants through a variety of physical, biological, and chemical treatment processes. As shown in the *Stormwater Technology Fact Sheet: Bioretention* (U.S. EPA, 1999d), Figure 11-1, displays runoff conveyed as sheet flow to the treatment area. This bioretention area consists of a grass buffer strip, sand bed, ponding area, organic layer or mulch layer, planting soil, and plants. The runoff's velocity is reduced by passing over or through a sand bed and subsequently distributed evenly along a ponding area. The ponding area, a surface organic layer and/or ground cover and the underlying planting soil, is graded and its center depressed. As the water gradually infiltrates the bioretention area or is evapotranspired, it is ponded to a depth of 15 cm (6 in.). Note that the bioretention area is graded to divert runoff in excess of the design storm volume. Exfiltration of the stored water in the bioretention area planting soil into the underlying soils occurs over a period of days.

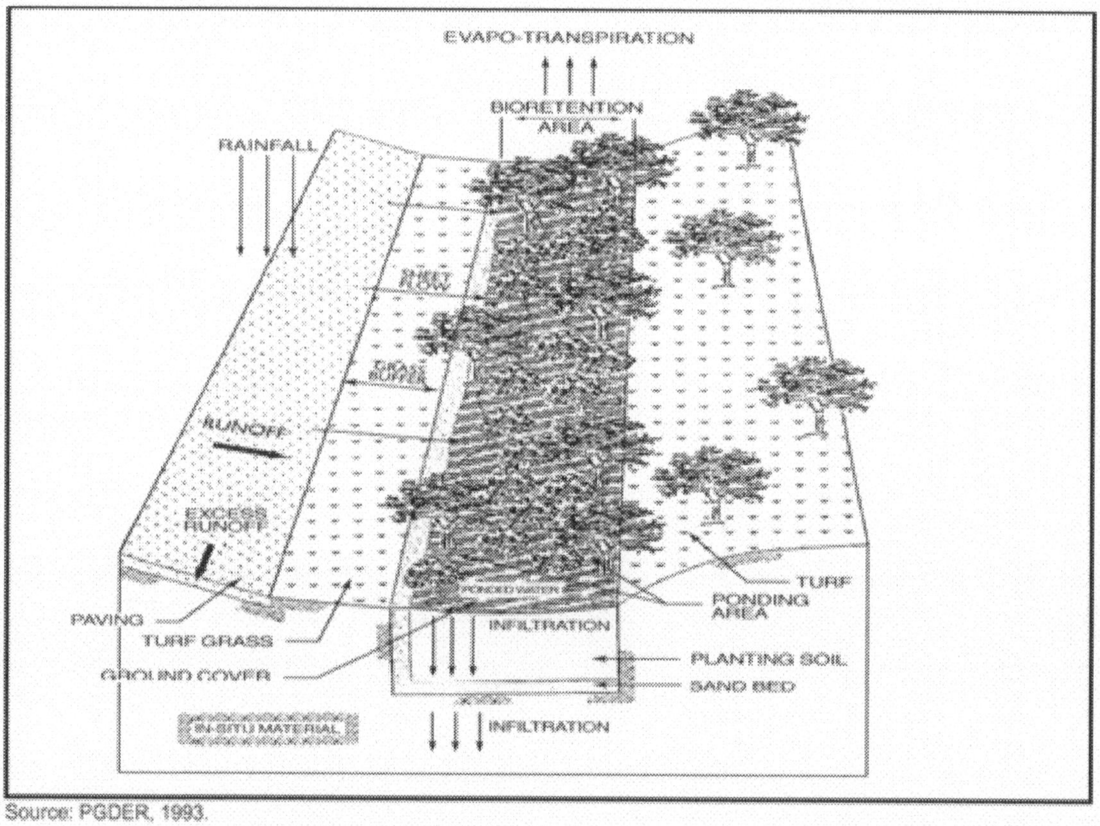

Source: PGDER, 1993.

Figure 11-1. Bioretention Area (U.S. EPA, 1999d).

Each component of the bioretention area is designed to perform a specific function as indicated in Table 11-1.

Table 11-1. Component Functions.

Design Criteria	Current Guidelines/Recommendations
Grass Buffer Strip	♦ Reduces runoff velocity and filters particulates from runoff.
Sand Bed	♦ Reduces velocity, filters particulates, and spreads flow over the length of the bioretention area. Aeration and drainage of the planting soil are provided by the 0.5 m (18 in.) deep sand bed.
Ponding Area	♦ Provides temporary storage location for runoff prior to its evaporation or infiltration. Particulates not filtered out by the grass filter strip or the sand bed settle within the ponding area.
Organic/Mulch Layer	♦ Filters pollutants and provides an environment conducive to the growth of microorganisms, which degrade petroleum-based products and other organic material. This layer acts in a similar way to the leaf litter in a forest and prevents the erosion and drying of underlying soils. The maximum sheet flow velocity prior to erosive conditions is 0.9 m/s (3 ft/s).
Planted Ground Cover	♦ Reduces the potential for erosion as well, slightly more effectively than mulch. The maximum sheet flow velocity prior to erosive conditions is 0.3 m/s (1 ft/s) for planted ground cover.
Clay in Planting Soil	♦ Provides adsorption sites for hydrocarbons, heavy metals, nutrients, and other pollutants.
Voids in Planting Soil	♦ Provides stormwater storage. The stored water and nutrients in the water and soil are then available to the plants for uptake.

The basic bioretention area can be customized to accommodate specific needs. For example, the City of Alexandria, Virginia, modified the bioretention BMP because impervious subsoils and marine clays prevented complete infiltration in the soil system. The redesigned bioretention BMP, which includes an underdrain within the sand bed that collects the infiltrated water and discharges it to a downstream sewer system, acts more as a filter that discharges treated water than as an infiltration device. Design modifications that will potentially include both aerobic and anaerobic zones in the treatment area are being reviewed. The anaerobic zone will promote denitrification.

A number of laboratory and field experiments have been conducted by the University of Maryland in conjunction with Prince George's County Department of Environmental Resources and the National Science Foundation in order to quantify the effectiveness of bioretention cells in terms of pollutant removal (Davis et al., 2001). A Website dedicated to this work can be found at http://www.ence.umd.edu/~apdavis/Bioret.htm.

11.2 Advantages

- Bioretention provides storm water treatment that enhances the quality of downstream water bodies by temporarily storing runoff in the BMP and releasing it over a period of four days to the receiving water (U.S. EPA, 1999d).
- The vegetation provides shade and windbreaks, absorbs noise, and improves an area's landscape.

11.3 Limitations

- The bioretention BMP is not recommended for areas with slopes greater than 20% or where removal of mature trees would be required because clogging may result, particularly if the BMP receives runoff with high sediment loads (U.S. EPA, 1999d).
- Bioretention is not a suitable BMP at locations where the water table is within 1.8 m (6 ft) of the ground surface and where the surrounding soil stratum is unstable.
- In cold climates the soil may freeze, preventing runoff from infiltrating into the planting soil.

11.4 Siting Criteria

Bioretention BMPs are generally used to treat stormwater from impervious surfaces at commercial, residential, and industrial areas (U.S. EPA, 1999d). Implementation of bioretention for stormwater management is ideal for median strips, parking lot islands, and swales. Moreover, the runoff in these areas can be designed to either divert directly into the bioretention area or convey into the bioretention area by a curb and gutter collection system.

The best location for bioretention areas is upland from inlets that receive sheet flow from graded areas and at areas that will be excavated (U.S. EPA, 1999d). In order to maximize treatment effectiveness, the site must be graded in a way that minimizes erosive conditions as sheet flow is conveyed to the treatment area. Locations where a bioretention area can be readily incorporated into the site plan without further environmental damage are preferred. Furthermore, to effectively minimize sediment loading in the treatment area, bioretention only should be used in stabilized drainage areas.

The layout of the bioretention area is determined after site constraints such as location of utilities, underlying soils, existing vegetation, and drainage are considered (U.S. EPA, 1999d). Sites with loamy sand soils are especially appropriate for bioretention because the excavated soil can be backfilled and used as the planting soil, thus eliminating the cost of importing planting soil.

The use of bioretention may not be feasible given an unstable surrounding soil stratum (e.g., Marlboro Clay), soils with a clay content greater than 25%, a site with slopes greater than 20%, and/or a site with mature trees that would be removed during construction of the BMP.

Bioretention can be designed to be offline or online of the existing drainage system (U.S. EPA, 1999d). The drainage area for a bioretention area should be between 0.1 and 0.4 ha (0.25 and 1.0 ac). Larger drainage areas may require multiple bioretention areas. Stabilized areas may erode when flow velocities are greater than 1.5 m/s (5 ft/s). In Maryland, such a flow generally occurs with a 10-year storm at 0.4 ha (1 ac) commercial or residential sites. The designer should determine the potential for erosive conditions at the site.

11.5 Design Guidelines

Prince George's County Department of Environmental Resources has specified bioretention design details in a document entitled *Design Manual for the Use of Bioretention in Storm Water Management* (PGDER, 1993). Design specifications were derived from research on soil adsorption capacities and rates, water balance, plant pollutant removal potential, plant adsorption capacities and rates, and maintenance requirements. A case study was performed using the specifications at three commercial sites and one residential site in Prince George's County, Maryland.

An example of sizing a facility is shown in Figure 11-2. For this discussion, sizing specifications are based on 1.3–1.8 cm of rainfall (0.5–0.7 in.) of precipitation over a 6-hour period (the mean storm event for the Baltimore-Washington area infiltrating into the bioretention area. Additional design guidelines are presented in Table 11-2.

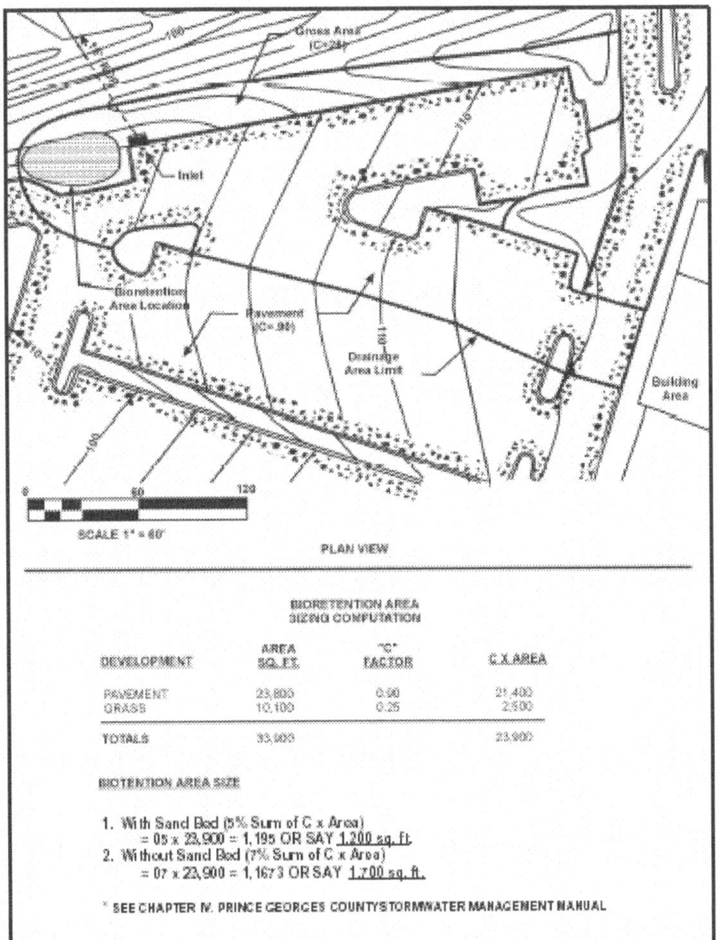

Figure 11-2. Example of Bioretention Sizing.

Table 11-2. Design Guidelines.

Design Criteria	Current Guidance/Recommendations
Hydraulic Sizing	♦ The size of the bioretention area, which is a function of the drainage area and the runoff generated from the area, should be 5%–7% of the drainage area multiplied by the rational method runoff coefficient, *c,* determined for the site. The 5% specification applies to a bioretention area that includes a sand bed and 7% applies to an area without one. ♦ To provide adequate storage and prevent water from standing for excessive periods of time the ponding depth of the bioretention area should not exceed 15 cm (6 in.). Water should not be left to stand for more than 72 hours. A restriction on the type of plants that can be used may be necessary due to some plants' water intolerance. Furthermore, if water is left is standing for longer than 72 hours mosquitoes and other insects may start to breed.
Planting Dimensions	♦ The *recommended* minimum dimensions of the bioretention area are 4.6 m (15 ft) wide by 12.2 m (40 ft) long, where the minimum width allows enough space for a dense, randomly distributed area of trees and shrubs to become established. This replicates a natural forest and creates a microclimate, thereby enabling the bioretention area to tolerate the effects of heat stress, acid rain, runoff pollutants, and insect and disease infestations that landscaped areas in urban settings typically are unable to tolerate. The *preferred* width is 7.6 m (25 ft), with a length of twice the width. Essentially, any facilities wider than 6.1 m (20 ft) should be twice as long as they are wide, which promotes the distribution of flow and decreases the chances of concentrated flow.
Soil	♦ The appropriate planting soil should be backfilled into the excavated bioretention area. Planting soils should be sandy loam, loamy sand, or loam texture with a clay content ranging from 10%–25%. ♦ Planting soil should be 10 cm (4 in.) deeper than the bottom of the largest root ball and 1.2 m (4 ft) altogether. This depth will provide adequate soil for the plants' root systems to become established, prevent plant damage due to severe wind, and provide adequate moisture capacity. Most sites will require excavation to obtain the recommended depth. ♦ Generally the soil should have infiltration rates greater than 1.2 cm (0.5 in.) per hour, which is typical of sandy loams, loamy sands, or loams. ♦ The pH of the soil should range between 5.5 and 6.5, where pollutants such as organic nitrogen and phosphorus can be adsorbed by the soil and microbial activity can flourish. Additional requirements for the planting soil include a 1.5%–3% organic content and a maximum 500 ppm concentration of soluble salts. Furthermore, the criteria for magnesium, phosphorus, and potassium are 39 kg/ac (35 lb/ac), 112 kg/ac (100 lb/ac), and 95 kg/ac (85 lb/ac), respectively. ♦ Soil tests should be performed for every 382 m^3 (500 yd^3) of planting soil, with the exception of pH and organic content tests, which are required only once per bioretention area (U.S. EPA, 1999d).
Plants	♦ Since high canopy trees may be destroyed during maintenance, the bioretention area should be vegetated to resemble a terrestrial forest community ecosystem that is dominated by understory trees. Also, the bioretention area should have discrete soil zones as well as a mature canopy and a distinct sub-canopy of understory trees, a shrub layer, and herbaceous ground covers. ♦ Three species each of both trees and shrubs are recommended to be planted at a rate of 2500 trees and shrubs per hectare (1000 per acre). For instance, a 4.6 m (15 ft) by 12.2 m (40 ft) bioretention area 55.8 m^2 (600 ft^2) would require 14 trees and shrubs. The shrub-to-tree ratio should be 2:1–3:1. On average, the trees should be spaced 3.6 m (12 ft) apart and the shrubs should be spaced 2.4 m (8 ft) apart. ♦ Native species tolerant to pollutant loads and varying wet and dry conditions should be used in the bioretention area. These species can be determined from several published sources, including *Native Trees, Shrubs, and Vines for Urban and Rural America*

Design Criteria	Current Guidance/Recommendations
	(Hightshoe, 1988). ♦ The designer should assess aesthetics, site layout, and maintenance requirements when selecting plant species. Adjacent non-native invasive species should be identified and the designer should take measures, such as providing a soil breach to eliminate the threat of these species invading the bioretention area. Regional landscaping manuals should be consulted to ensure that the planting of the bioretention area meets the landscaping requirements established by the local authorities.
	♦ The designers should evaluate the best placement of vegetation within the bioretention area. Plants should be placed at irregular intervals to replicate a natural forest. Trees should be placed on the perimeter of the area to provide shade and shelter from the wind. Trees and shrubs can be sheltered from damaging flows if they are placed away from the path of the incoming runoff. In cold climates, species that are more tolerant to cold winds, such as evergreens, should be placed in windier areas of the site.
Final Cover	♦ Following placement of the trees and shrubs, the ground cover and/or mulch should be established. Mulch depths should be kept below 7.6 cm (3 in.) because more would interfere with the cycling of carbon dioxide and oxygen between the soil and the atmosphere. Ideally mulch should be aged for 1 year (minimum of at least 6 months) and applied uniformly over the site.

11.6 Construction Guidelines

- Planting soil depths of greater than 1.2 m (4 ft) may require additional construction practices such as shoring measures (U.S. EPA, 1999d). Planting soil should be placed in 7 cm (8 in.) or greater lifts and lightly compacted until the desired depth is reached.
- In the metropolitan Washington, D.C., area, trees, and shrubs should be planted from mid-March through the end of June or from mid-September through mid-November (U.S. EPA, 1999d). Planting periods in other areas of the United States vary. Vegetation should be watered at the end of each day for 14 days following planting.
- Ground cover such as grasses or legumes can be planted during the spring of the year. Mulch should be placed immediately after trees and shrubs are planted. From 5 to 8 cm (2 to 3 in.) of commercially available, fine-shredded hardwood mulch or shredded hardwood chips should be applied to the bioretention area to protect from erosion.
- If possible, divert runoff (other than necessary irrigation) during the period of vegetation establishment.

11.7 Performance

Bioretention removes storm water pollutants through physical and biological processes, including adsorption, filtration, plant uptake, microbial activity, decomposition, sedimentation, and volatilization (U.S. EPA, 1999d). Adsorption is the process whereby particulate pollutants attach to soil (e.g., clay) or vegetation surfaces. Adequate contact time between the surface and pollutant must be provided for in the design of the system for this removal process to occur. Thus, the infiltration rate of the soils must not exceed those specified in the design criteria or pollutant removal may decrease. Pollutants removed by adsorption include metals, phosphorus, and hydrocarbons. Filtration occurs as runoff passes through the bioretention area media, such as the sand bed, ground cover, and planting soil.

Common particulates removed from stormwater include particulate organic matter, phosphorus, and suspended solids. Biological processes that occur in wetlands result in pollutant uptake by plants and microorganisms in the soil. Plant growth is sustained by the uptake of nutrients from the soils, with woody plants locking up these nutrients through the seasons. Microbial activity within the soil also contributes to the removal of nitrogen and organic matter. Nitrogen is removed by nitrifying and denitrifying bacteria, while aerobic bacteria are responsible for the decomposition of the organic matter. Microbial processes require oxygen and can result in depleted oxygen levels if the bioretention area is not adequately aerated. Sedimentation occurs in the swale or ponding area as the velocity slows and solids fall out of suspension. The removal effectiveness of bioretention has been studied during field and laboratory studies conducted by the University of Maryland (Davis et al., 1998). During these experiments, synthetic stormwater runoff was pumped through several laboratory and field bioretention areas to simulate typical storm events in Prince George's County, Maryland. Removal rates for heavy metals and nutrients are shown in Table 11-3.

Table 11-3. Laboratory and Estimated Bioretention (Davis et al., 1998; PGDER, 1993).

Pollutant	Removal Rate
Total Phosphorus	70%–83%
Metals (Cu, Zn, Pb)	93%–98%
TKN	68%–80%
Total Suspended Solids	90%
Organics	90%
Bacteria	90%

Results for both the laboratory and field experiments were similar for each of the pollutants analyzed. Doubling or halving the influent pollutant levels had little effect on the effluent pollutants concentrations (Davis et al., 1998). The microbial activity and plant uptake occurring in the bioretention area will likely result in higher removal rates than those determined for infiltration BMPs.

11.8 Maintenance

The primary maintenance requirement for bioretention areas is that of inspection and repair or replacement of the treatment area's components. Generally, this involves nothing more than the routine periodic maintenance that is required of any landscaped area. Native plant species should have been selected for use in the bioretention cell, reducing fertilizer, pesticide, water, and overall maintenance requirements. Bioretention system components should blend over time through plant and root growth, organic decomposition, and the development of a natural soil horizon. These biologic and physical processes over time will lengthen the facility's life span and reduce the need for extensive maintenance. Maintenance requirements are summarized in Table 11-4.

Table 11-4. Typical Maintenance for Bioretention Areas.

Activity	Schedule
♦ Evaluate the trees and shrubs and subsequently remove any dead or diseased vegetation (U.S. EPA, 1999d). Diseased vegetation should be treated as needed using preventative and low-toxic measures to the extent possible.	Biannually
♦ Replace mulch when erosion is evident or when the site begins to look unattractive. Specifically, the entire area may require mulch replacement every two to three years, although spot mulching may be sufficient when there are random void areas. Mulch replacement should be done during the spring and the old mulch should be removed prior to distribution of the new mulch.	As necessary or every 2–3 years
♦ Apply an alkaline product, such as limestone, to counteract soil acidity resulting from slightly acidic precipitation and runoff. Before the limestone is applied, the soils and organic layer should be tested to determine the pH and therefore the quantity of limestone required. When levels of pollutants reach toxic levels which impair plant growth and the effectiveness of the BMP, soil replacement may be required (PGDER, 1993).	1–2 times per year
♦ New Jersey's Department of Environmental Protection states in their bioretention systems standards that accumulated sediment and debris removal (especially at the inflow point) will normally be the primary maintenance function. Other potential tasks include replacement of dead vegetation, soil pH regulation, erosion repair at inflow points, mulch replenishment, unclogging the underdrain, and repairing overflow structures.	As needed
♦ Replace soils if cation exchange capacity is significantly reduced.	Possibly every 5–10 years

11.9 Costs

11.9.1 Construction Costs

Construction cost estimates for a bioretention area are slightly greater than those for the required landscaping for a new development (U.S. EPA, 1999d). A general rule of thumb (Coffman, Goo, and Frederick, 1999) is that residential bioretention areas average about \$32to \$43 per m^2 (\$3 to \$4 per ft^2), depending on soil conditions and the density and types of plants used. Commercial, industrial and institutional site costs can range between \$110 and \$430 per m^2 (\$10 to \$40 per ft^2), based on the need for control structures, curbing, storm drains, and underdrains.

Bioretention areas recently constructed in Prince George's County, Maryland, cost approximately \$500 (U.S. EPA, 1999d). These units are rather small (400 ft^2 or 37 m^2) and their cost relatively low. The cost estimate includes the cost of excavating 0.6–1 m (2–3 ft) and vegetating the site with one to two trees and three to five shrubs. The cost of planting soil, which is not included in the estimate, increases the overall cost for a bioretention area. Retrofitting a site typically costs more, averaging \$6500 per bioretention area. The higher costs are attributed to the demolition of existing concrete, asphalt, and existing structures and the replacement of fill material with planting soil. The costs of retrofitting a commercial site in Maryland, Kettering Development, with 15 bioretention areas were estimated at \$111,600.

In any bioretention area design, the cost of plants varies substantially and can account for a significant portion of the expenditures. While these cost estimates are slightly greater than those of typical landscaping treatment (due to the increased number of plantings, additional soil excavation, backfill material, use of underdrains, and so forth), those landscaping expenses that would be required regardless of the bioretention installation should be subtracted when determining the net cost.

Perhaps of most importance, however, the cost savings compared to the use of traditional structural stormwater conveyance systems makes bioretention areas quite attractive financially. For example, the use of bioretention can decrease the cost required for constructing storm water conveyance systems at a site. A medical office building in Maryland was able to reduce the amount of storm drain pipe that was needed from 240 to 70 m (800 to 230 ft)—a cost savings of $24,000 (PGDER, 1993). And a new residential development spent a total of approximately $100,000 using bioretention cells on each lot instead of nearly $400,000 for the traditional stormwater ponds that were originally planned. Also, in residential areas, stormwater management controls become a part of each property owner's landscape, reducing the public burden to maintain large centralized facilities.

11.9.2 Maintenance Costs

The operation and maintenance costs for a bioretention facility are comparable to those of typical landscaping required for a site. Costs beyond the normal landscaping fees include the cost of soil testing and may include costs for a sand bed and planting soil.

11.10 Research Needs

Bioretention is a relatively new, onsite stormwater treatment system; consequently, there are significant uncertainties about water quality performance and long-term maintenance requirements to prevent clogging. Although there are limited data on construction costs, bioretention systems consist mainly of landscaping, and those costs can be estimated quite easily.

11.11 Additional Resources

Growing Greener in your Rappahannock River Watershed http://for.communitypoint.org/pages/grogreen.html.

Low Impact Development (LID) Center, 2003 (revised), Low impact development (LID) urban design tools. Beltsville, MD. [http://www.lid-stormwater.net/]

Prince George's County Department of Environmental Resources (PGDER), 1997, *Bioretention Monitoring: Preliminary Data Analysis.* Division of Environmental Management, Watershed Protection Branch. Landover, MD.

Prince George's County Department of Environmental Resources (PGDER), 2001, *The Bioretention Manual.* Programs and Planning Division, MD.

U.S. Environmental Protection Agency (U.S. EPA), 2000, *Low Impact Development (LID): A Literature Review.* EPA-841-B-00-005.United States Environmental Protection Agency, Office of Water, Washington, DC.

CHAPTER 12.0

POROUS/PERVIOUS PAVING

12.1 Description

Pervious paving as used in the U.K. is used for light vehicle loading in car parks and accesses. The term describes a system comprising a load-bearing, durable surface together with an underlying layered structure that temporarily stores water prior to infiltration or drainage to a controlled outlet. A photograph of a typical porous pavement installation is shown on Figure 12-1, and a cross-section is shown on Figure 2-2. The surface can itself be porous such that water infiltrates across the entire surface of the material (e.g., grass and gravel surfaces, porous concrete and porous asphalt), or can be built up of impermeable blocks separated by spaces and joints, through which the water can drain. This latter system is termed 'permeable' paving. Attenuation of flow is provided by the storage within the underlying structure or sub base, together with appropriate flow controls. An underlying geotextile may permit groundwater recharge, thus contributing to the restoration of the natural water cycle. Alternatively, where infiltration is inappropriate (e.g., if the groundwater vulnerability is high, or the soil type is unsuitable), the surface can be constructed above an impermeable membrane. The system offers a valuable solution for drainage of spatially constrained urban areas.

Figure 12-1. Typical Porous Pavement Installation.

Significant attenuation and improvement in water quality can be achieved by permeable pavements, whichever method is used. The surface and subsurface infrastructure can remove both the soluble and fine particulate pollutants that occur within urban runoff. Roof water can be piped into the storage area directly, adding areas from which the flow can be attenuated. Also, within lined systems, there is the opportunity for stored runoff to be piped out for reuse.

Constructed pervious surfaces can remove or reduce the requirement for conventional drainage networks and ancillaries such as pipes, gullypots, manholes, and interceptors.

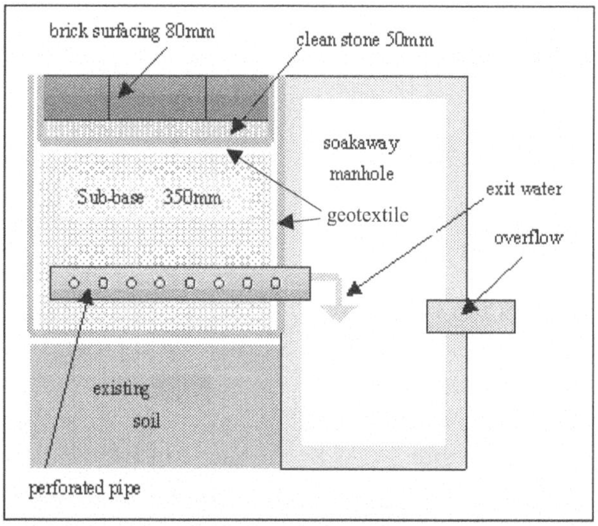

Figure 12-2. Typical Permeable Paving Cross-Section and Outlet Detail.

Depending on the surface material and the use of the surface, the underlying construction may be:

- Soil, sharp sand, crushed rock; and
- One of more layers of different materials (e.g., granular and/or geosynthetic layers)

It may also include a permeable geotextile or an impervious geomembrane and, if necessary, drainage networks to discharge water entering or flowing through construction. Outflow options are shown on Figure 12-3 (CIRIA, 2002).

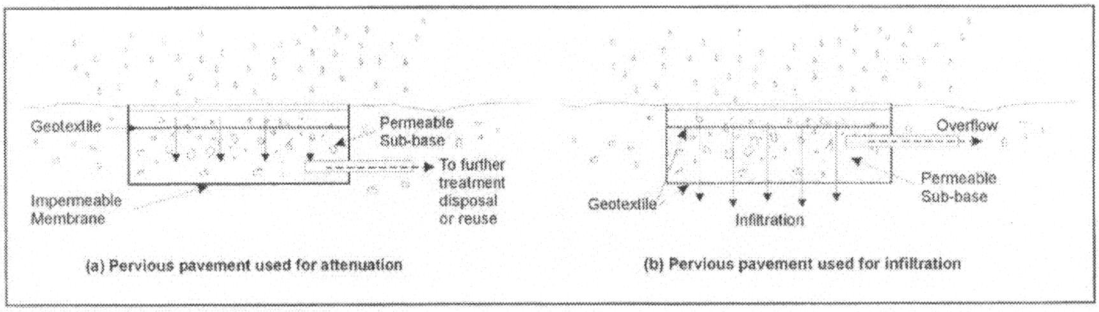

Figure 12-3. Outflow Options (CIRIA, 2002).

A large number of permeable paving systems had been installed in Scotland up to December 2001. The recently compiled SEPA database (Wild, Jeffries, and D'Arcy, 2002) shows that some 25% of all SUDS sites in Scotland (approximately 150) include permeable paving.

Most installations are in low trafficked areas with consequent low pollutant levels. The oldest sites are up to six years old and to date have had little or no maintenance except where necessitated by some construction defect or error. Early versions had porous blocks, which failed due to poor crush resistance. Solid blocks are now used which are not themselves porous but provide vertical drainage channels at each block. A permeable block detail is shown on Figure 12-4. Additionally, principal access routes in car parks now use conventional road construction and the runoff from these accesses is directed onto the permeable paving "zone" for flow and pollutant attenuation.

Figure 12-4. Permeable Block Detail (From Dierkes et al., 2002).

The function of porous paving in the treatment train is to provide source control of flow and pollutants within car parking areas. This is a particular advantage where the car parking is in areas where there may not be other significant pollution sources—for example at offices in otherwise residential areas. The small block construction of proprietary paving systems suits commercial and retail developments because changes to parking layouts are easily undertaken.

Clogging can be expected at specific locations and vacuum cleaning vehicles have been developed, which have been shown to restore infiltration rates in blocked systems.

As the SUD system is totally enclosed, there are minimal safety risks associated with its operation and use. Indeed, the improved skid resistance associated with the blockwork surface may improve safety in winter conditions.

12.2 Advantages

- Reduces runoff volume, while providing stormwater treatment; and
- Unobtrusive, resulting in high level of acceptability.

12.3 Limitations

The use of permeable pavement may be restricted in cold regions, arid regions, or regions with high wind erosion. There are some specific disadvantages associated with permeable pavement, which are as follows:

- Permeable pavement can become clogged if improperly installed or maintained. However, this is countered by the ease with which small areas of paving can be cleaned or replaced when blocked or damaged.
- The application should be limited to highways with low traffic volumes, axle loads and speeds (less than 30 mph limit), car parking areas and other lightly trafficked or nontrafficked areas. Permeable surfaces are currently not considered suitable for more heavily trafficked roads, due to the risks associated with failure on high-speed roads, the safety implications of ponding, and disruption arising from reconstruction.
- When using unlined, infiltration systems, there is some risk of contaminating groundwater, depending on soil conditions and aquifer susceptibility. However, this risk is likely to be small, because the areas drained tend to have inherently low pollutant loadings.
- The use of permeable pavement is restricted to gentle slopes, so car parking tends to be terraced.
- Porous block paving has a higher risk of abrasion and damage than solid blocks.

12.4 Siting Criteria

If the grades, subsoils, drainage characteristics, and groundwater conditions are suitable, permeable paving may be substituted for conventional pavement on parking areas, cul de sacs, and other areas with light traffic. Slopes should be flat or very gentle. Scottish experience has shown that permeable paving systems can be installed in a wide range of ground conditions, and the flow attenuation performance is excellent even when the systems are lined. The suitability of a pervious system at a particular pavement site will, however, depend on the loading criteria required of the pavement.

Where the system is to be used for infiltrating drainage waters into the ground, the vulnerability of local groundwater sources to pollution from the site should be low, and the seasonal high-water table should be at least 4 ft below the surface.

Ideally, the pervious surface should be horizontal in order to intercept local rainfall. On sloping sites, pervious surfaces may be terraced to accommodate differences in levels.

12.5 Design Guidelines

There are three elements to be considered in the design of pervious surfaces subject to vehicular traffic:

- The surface course;
- The underlying construction layers; and
- The foundation soils.

For specific design details, refer to the CIRIA guidance RP582 *Source Control Using Constructed Pervious Surfaces* (2002). Key points are summarized in Table 12-1.

Table 12-1. Design Guidelines.

Design Criteria	Current Guidance/Recommendations
General	♦ The design of each layer of the pavement must be determined by the likely traffic loadings and the required operational life. To provide satisfactory performance, the following criteria must be met: ♦ The subgrade must be able to sustain traffic loading without excessive deformation. ♦ The granular capping and sub-base layers must give sufficient load-bearing to provide an adequate construction platform and base for the overlying pavement layers. ♦ The pavement materials must not crack or suffer excessive rutting under the influence of traffic. This is controlled by the horizontal tensile stress at the base of these layers.
Pavement Layer (Surface Course)	♦ There is no current structural design method in the United Kingdom specifically for pervious pavements. Allowances must be made for the following factors in the design and specification of materials: ♦ Pervious pavements use materials with high permeability and void space. All the current U.K. pavement design methods are based on the use of conventional materials that are dense and relatively impermeable. The stiffness of the materials must therefore be assessed. ♦ Water is present within the construction and can soften and weaken materials, and this must be allowed for. ♦ Existing design methods assume full friction between layers. Any geotextiles or geomembranes must be carefully specified to minimize loss of friction between layers. ♦ Porous asphalt loses adhesion and becomes brittle as air passes through the voids. Its durability is therefore lower than conventional materials. ♦ The single-sized grading of materials used means that care must be taken to ensure that loss of finer particles between unbound layers does not occur.
Surface Substructure	♦ Positioning a geotextile near the surface of the pervious construction should enable pollutants to be trapped and retained close to the surface of the construction. This has both advantages and disadvantages. The main disadvantage is that the filtering of sediments and their associated pollutants at this level may hamper percolation of waters and can eventually lead to surface ponding. One advantage is that even if eventual maintenance is required to reinstate infiltration, only a limited amount of the construction needs to be disturbed, since the sub-base below the geotextile is protected. In addition, the pollutant concentration at a high level in the structure allows for its release over time. It is slowly transported in the stormwater to lower levels where chemical and biological processes may be operating to retain or degrade pollutants. ♦ The design should ensure that sufficient void space exists for the storage of sediments to limit the period between remedial works. ♦ Pervious pavements require a single-size grading to give open voids. The choice of materials is therefore a compromise between stiffness, permeability, and storage capacity. ♦ Because the sub-base and capping will be in contact with water for a large part of the time, the strength and durability of the aggregate particles when saturated and subjected to wetting and drying should be assessed. ♦ A uniformly graded single-size material cannot be compacted and is liable to move when construction traffic passes over it. This effect can be reduced by the use of angular crushed rock material with a high surface friction. ♦ If an approach similar to that taken in the United States is followed in the United Kingdom, then it is possible to use aggregate in accordance with BS 882 (1992). The most appropriate are the 40 mm (1.6 in.) to 5 mm (0.20 in.) and 20 mm (0.79 in.) to 5 mm (0.20 in.) crushed rock coarse aggregate.

Design Criteria	Current Guidance/Recommendations
Granular Sub-base	♦ In pollution-control terms, these layers represent the site of long-term chemical and biological pollutant retention and degradation processes. The construction materials should be selected, in addition to their structural strength properties, for their ability to sustain such processes. In general, this means that materials should create neutral or slightly alkaline conditions, and they should provide favorable sites for colonization by microbial populations.
Frost Resistance/Freeze-Thaw of Saturated Layers	♦ Materials within 450 mm of the surface should not be frost susceptible, unless it is proved that they can limit frost penetration to a shallower depth. ♦ The sub-base material should not become fully saturated. To prevent this, and thereby leave space for water volume expansion on freezing, the design storage volume should be increased by 30%. ♦ Deicing salts and grit may interfere with the operation of the device.
Geotextiles and Geogrids	♦ To reduce the effects of water on the strength and stiffness properties of the subgrade below pervious pavements, a geotextile separator layer should always be provided over moisture-sensitive, fine-grained subgrades. This will enhance the performance of the pavement structure in two ways: (1) preventing pumping of finer soils into the open graded aggregate of the unbound layers; and (2) enhancing resistance to deformation of the subgrade under load, by providing a tensile resistance. ♦ Geogrids may also be incorporated into the granular layers in order to increase the stiffness and reduce the required thickness of the layers. They can also be used as separation layers to prevent migration of fine particles between unbound layers, and play an important role in removing pollutants. ♦ The choice of geotextile should be based on the required performance characteristics, including strength, puncture resistance and the criteria for filtration.
Geomembranes	♦ If an impermeable geomembrane is required to prevent water infiltration, create attenuation, or prevent softening of the subgrade, it should be able to withstand the loads imposed during construction and the service life of the pavement. ♦ It should be manufactured from a robust material such as polypropylene or high-density polyethylene (HDPE). ♦ In order to ensure that it is impermeable, it should have homogeneous welded joints and be installed to ensure that no water can penetrate into the subgrade. In addition, it is good practice to protect the geomembrane from damage on both faces with a geotextile fleece or sand layer.
Trees	♦ The effects of prolonged exposure to wetting/drying cycles will also need to be considered, particularly for plastic clays near trees. In these situations the clay dries and shrinks when trees remove the moisture and swells again when moisture is added. To reduce the effects of trees, either greater than normal falls in subgrade levels and pipes should be allowed or trees should be planted where they will not influence the soils below the pavement.
Infiltration	♦ To allow infiltration of the surface water into the subgrade, the soils should be of sufficient permeability and a suitable depth above the water table. ♦ Infiltration systems should be designed in accordance with CIRIA Report 156 (CIRIA, 1996).
Storage	♦ The system should be designed so that the full design rainfall event can be stored, assuming a constrained maximum discharge rate, as required by the environmental regulator for the site.

12.6 Construction Guidelines

- There should be no ancillary drainage structures, such as manholes and gullypots.
- Permeable surfaces can be laid without cross-falls or longitudinal gradients.
- The blocks should be level.

- They should not be used for storage of site materials, unless the surface is well protected from deposition of silt and other spillage.
- The pavement should be constructed in a single operation, as one of the last items to be built, on a development site. Landscape development should be completed before pavement construction to avoid contamination by silt or soil from this source.
- Surfaces draining to the pavement should be stabilized before construction of the pavement.
- Inappropriate construction equipment should be kept away from the pavement to prevent damage to the surface, sub-base, or subgrade.

12.7 Performance

An extensive review of the performance of permeable paving systems was written by Pratt (2001). This gives a very detailed review of studies to date.

12.7.1 Hydraulic Performance

All the performance data reviewed by Pratt (2001) very clearly show significant attenuation of flow by virtue of the storage provided.

Laboratory studies in Nottingham were undertaken on a small, permeable-block-surfaced parking area (Pratt, Mantle, and Schofield, 1990). The investigators measured considerable reduction in peak discharges, and a reduction of the volume of stormwater discharged during the period of rainfall to about 40% of the total discharge, as compared with the performance of traditional impermeable surfaces, which discharge close to 100% of runoff within the storm duration.

Similar observations were made at Wheatley Motorway Service Area (Abbott, Comino, and Angood, 2000), where there is a 6250-m^2 porous-block-surfaced car parking area. The volume of rainfall draining from the sub-base during rainfall events varied from 4% to 47% of the rainfall volume, with an average value of 22.5%. Some events took two to three days to outflow after rainfall ceased, producing the impression of a system base flow. The mean event percentage runoff measured was 67%. This indicates that evaporation from permeable pavements is significant—up to 30% being lost from below the surface. The peak outflow is markedly reduced, a peak rainfall intensity of 12 mm/h limited to 0.4 mm/h at the outflow.

The attenuation function of pervious systems is further illustrated in Figure 12-5, using data from a study at the Royal Bank of Scotland headquarters in Edinburgh (Schlüter and Jefferies, 2001). It will be seen that the attenuation may be expressed as a function of the rainfall total and is more significant for smaller rainfall events. The data shown in Figure 12-5 were obtained from a lined system—one with an underlying synthetic impermeable barrier—from which there is no infiltration into the underlying deposits.

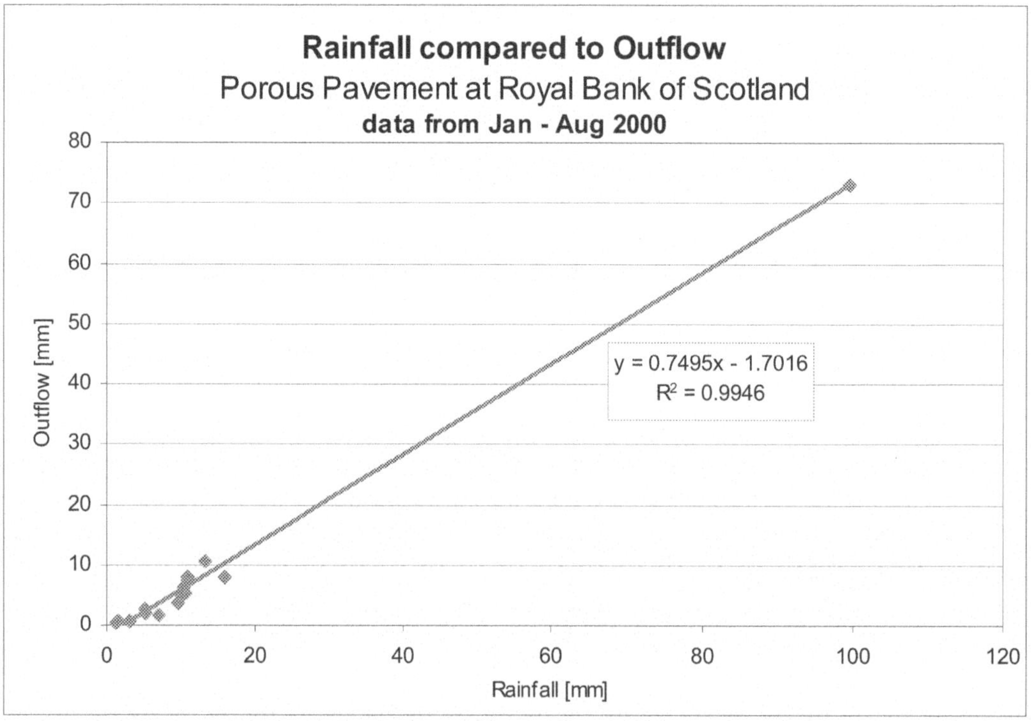

Figure 12-5. Percentage Runoff Data from Permeable Pavement (Schlüter and Jefferies, 2001).

Comparisons of monitoring results from porous and impermeable paving studies show the significant attenuation provided by porous arrangements. Key results are shown in Table 12-2 and Figure 12-6 through Figure 12-8, which are taken from Macdonald and Jefferies (2001). Figure 12-6 compares the runoff from adjacent systems. The delay of approximately 8 hours before outflow from the porous system is clearly illustrated, as is the smaller total volume of runoff from the porous system.

Table 12-2. Key Hydraulic Parameters for NATS Car Park (Macdonald and Jefferies, 2001).

	Percentage Outflow			**Lag Time**		**Peak Flow Reduction**			**mm Rain Before Runoff**	
	Tarmac	Porous	**Benefit Factor**	Tarmac	Porous	Tarmac	Porous	% Reduction	Tarmac	Porous
Min	21.4	2.5	6.5	-158	29	0.93	0.05	23.7	0	3.6
Max	72.8	66	91.2	123	600	17	3.3	98.4	2.6	18.6
Mean	**48.2**	**21.7**	**64.2**	**9.3**	**181.5**	**5.17**	**1.1**	**76.8**	**0.76**	**7.4**

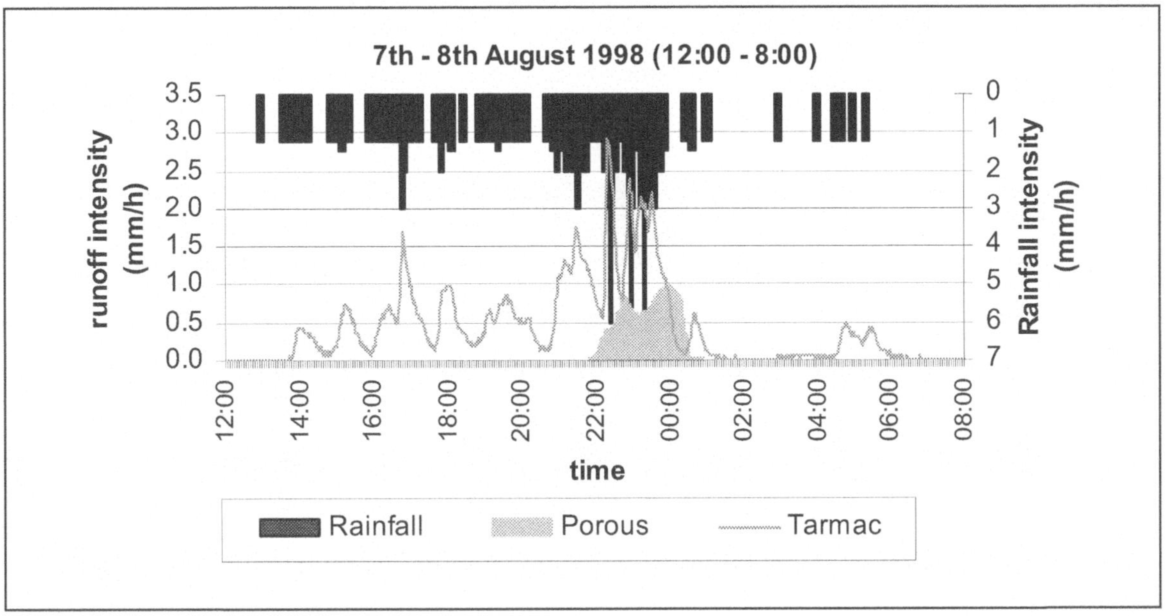

Figure 12-6. Comparison of Runoff from Adjacent Porous and Impermeable Paving (Macdonald and Jefferies, 2001).

Figure 12-7 shows the effect of antecedent rainfall on the two adjacent car park areas, API5 being a factor which indicates the depth of rainfall in the days prior to the event in question. Very little rain is needed to produce rainfall from the impermeable tarmac car park, whereas a minimum of 3.5 mm (0.14 in.) and as much as 19 mm (0.75 in.) of rainfall can fall before runoff occurs from a porous system.

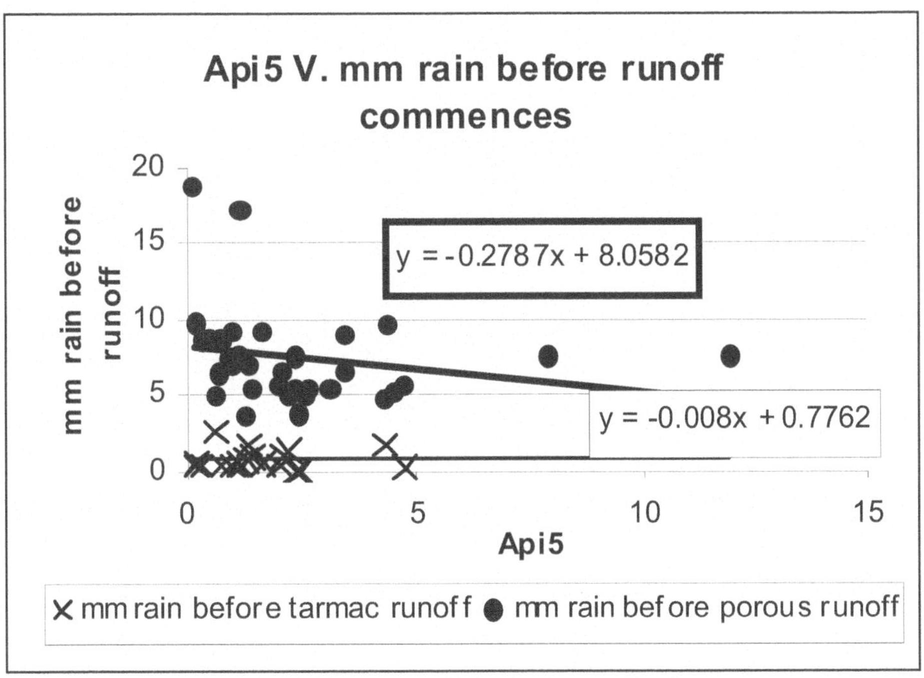

Figure 12-7. Effect of Antecedent Rainfall on Porous and Impermeable Pavements (Macdonald and Jefferies, 2001).

Figure 12-8 shows the reduction in ponding and surface water runoff from a permeable system, when compared to an equivalent impermeable asphalt surface close by.

Figure 12-8. Visual Comparison of Porous and Impermeable Pavements at the Same Site in Edinburgh (Macdonald and Jefferies, 2001).

12.7.2. Infiltration and Blockage

Recently developed porous blocks in the United Kingdom have a through flow rate of 3500 mm/hr (140 in./hr) individually and an as-laid rate of 4500 mm/hr (180 in./hr) when joints also allow through-flow (Abbot, Comino, and Angood, 2000). For porous macadam, Pratt measured rates of average 39,000 mm/hr (1500 in./hr) on a five-year old surface in Nottingham (Pratt, 1995).

Water quality research has indicated TSS removals between 40% and 100%. However, the more sediment trapped by the pavement, the sooner the pavement is likely to become blocked and require rehabilitation.

Abbot, Comino, and Angood (2000) reported rapid deterioration (10 months) in the surface infiltration rate of the porous bricks. This was due to widespread clogging of the surface of the car park. However, infiltration through the gaps between blocks remained high (up to 10 times higher).

The problems with blockage and difficulties of cleansing to reinstate infiltration rates led to the assessment by Pratt, Newman, and Bond (1998) of block paving. A laboratory model of a permeable pavement, including concrete block, gravel, and geotextile was tested to monitor the deposition of sediments as various levels below the surface. After the equivalent of 10 years' rainfall, the model was taken apart. The model suggested that sediment was transported with stormwater inflow throughout the surface structure, and failure of the stormwater to infiltrate the surface could not be assumed to be caused by simple blockage of the holes, but rather the general filling of the gravel bedding and the inlet holes—leading to collection of sediment on the surface. Prior to failure, the accumulation of sediment on the gravel layer would impede movement into the sub-base—increasing pollutant concentrations at the surface.

12.7.3 Water Quality Performance

In general, studies have indicated that permeable pavements:

- Are effective at filtering sediments from stormwater, which may affect infiltration rates at the surface in time;
- Provide internal sites for the retention and adsorption of pollutants, such as oils and heavy metals, which may accumulate steadily over time and require careful removal, when the pavement is reconstructed; and
- May be used as an aerobic bioreactor to degrade oils and grease, given appropriate nutrients and ambient conditions to maintain a viable microbial population (Pratt, 1990).

SEPA reports that removal of up to 80% of sediment, 60% of phosphorus, and 80% of nitrogen has been measured, as well as high removal rates for trace metals and organic matter. These rates are all slightly higher than those in Table 12-3, which summarizes water quality results from a number of studies. A graphical representation of those results is shown in Figure 12-9.

Table 12-3. Porous Paving — Comparison of Water Quality Performance from Different Studies.

Permeable Pavements — Water Quality Performance — Average % Removal													
SITE		**Zn**	**Pb**	**Cd**	**Ni**	**Cr**	**Cu**	**SS**	**NH3**	**NO3**	**HC**	**COD**	**o-phos**
NATS Formpave (Macdonald and Jefferies, 2001)	%	42	66	3	63	26	-580	32	32	32	69	49	-159
Number of events		6	6	6	6	6	6	6	6	6	7	6	6
Rue Classerie , Nantes. Pervious Asphalt (Legret and Colandini, 1998)	%	73	84	77			0	59					
Number of events		11	11	11			11	11					
Car Park (Sweden) (Raimbault, 1997)	%		40					95					
Number of events			?					?					
Park and Ride (France) (Raimbault, 1997)	%		86					36				79	
Number of events			?					?				?	
Car Park (France) (Raimbault, 1997)	%		85					94				77	
Number of events			?					?				?	
Street (France) (Raimbault, 1997)	%		79					64					
Number of events			?					?					

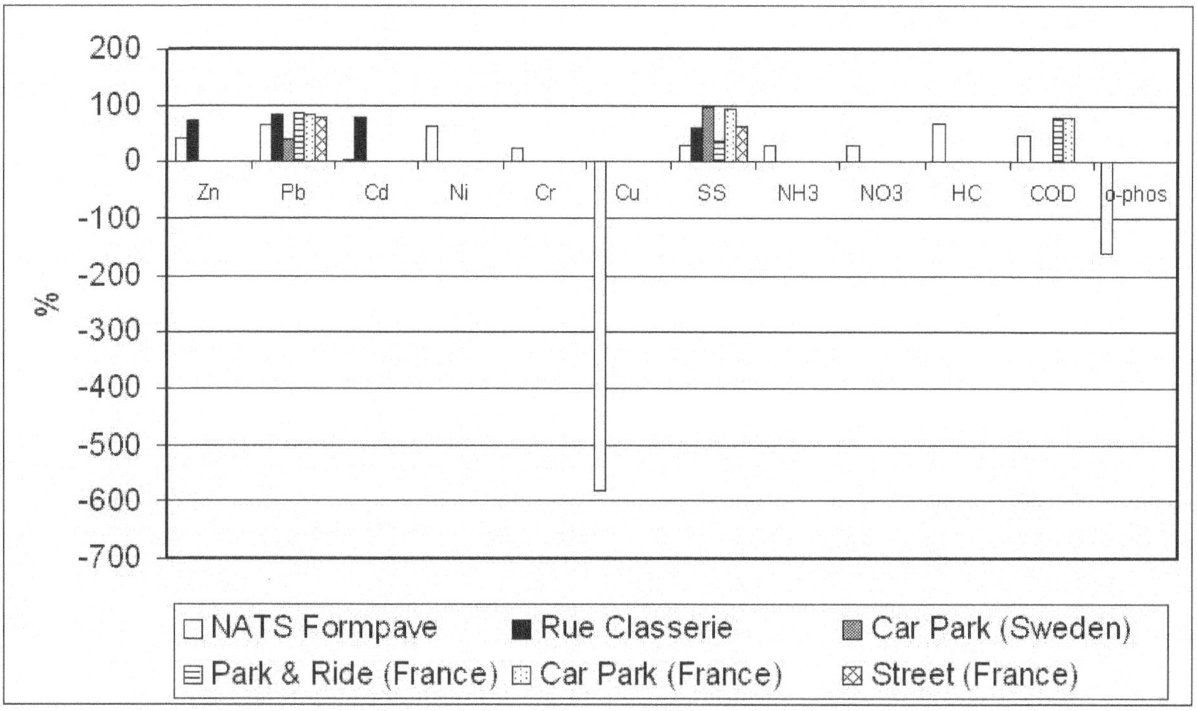

Figure 12-9. Water Quality Performance.

Macdonald and Jefferies (2001) proposed that the N, NH_3, and orthophosphate increases at the NATS site may be due to bacterial biofilm in the sub-base, which would produce ammonium and other nutrients from decaying plants and bacteria.

Table 12-4 and Table 12-5 present more detailed data summarized from Macdonald and Jefferies (2001) from two studies at a car park at National Air Traffic Services (NATS) in Edinburgh.

Table 12-4. Water Quality Parameters — from Samples Collected at NATS (Macdonald and Jefferies, 2001).

Parameter		pH	Cond.	TSS	BOD	TON	o-phos	Chlor.	Amm N
Unit			µS/cm	mg/L	mg/L	mg/L	mg/L	mg/L	mg/L
Average	Tarmac	6.7	49	35	4.3	0.7	0.03	8.3	0.2
	Porous	8	315	17	1.9	0.9	0.15	24.4	0.34
Range of EMCs	Tarmac	6.4–6.85	41-62	15.8-51	2.8-5.8	0.15-1.42	0.02-0.04	6.5-10.9	0.03-0.48
	Porous	7.73-8.2	210-416	9.8-24	1.2-2.6	0.36-1.36	0.05-0.53	5.1-57.6	0.03-1.13
EMC Red./ Inc. (-/+)		+1.4	+822%	-32%	-49%	+165%	+159%	+389%	-32%

Table 12-5. NATS Outflow Key Heavy Metals Concentrations.

		Cd	Pb	Cr	Cu	Ni	Zn	Hydrocarbons
		µg/L	µg/L	µg/L	µg/L	µg/L	µg/L	mg/L
Outflow NATS (EMC)	Min	0.12	0.93	3.85	3.76	0.95	17	0.15
	Max	5.33	24.3	8.73	23.07	8.69	67	0.47
	Mean	1.91	9.8	5.73	10.9	3.78	42	1.21
EQS 2 *	Max	5	50	150	1	50	75	-
Service Reservoir	Mean	<0.1	7.6	<1.7	23	0.6	8.2	-

* EQS = Environmental Quality Standards

Recently, Raimbault 1997 has examined samples of porous asphalt concrete pavements to identify the location of pollutants within the construction. The results of this analysis are shown on Figure 12-10. The measured heavy metal concentrations show that micro pollutants are mostly retained at the surface of the porous asphalt, with a slight increase of heavy metals concentration at the geotextile and at the upper layers of the soil underneath the reservoir. However, after four and eight years of operation, the levels remained lower than those of polluted soils elsewhere.

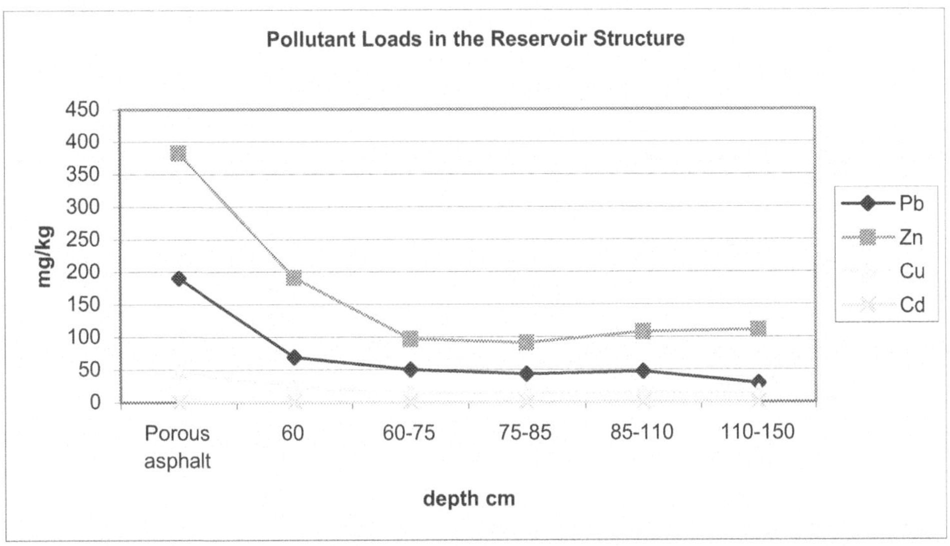

Figure 12-10. Pollutant loadings (Raimbault, 1997).

German field studies have shown excellent pollution retention capabilities, systems which incorporate vegetation showing better performance (Dierkes et al., 2002). This study also showed a slight increase of heavy metals concentrations in the top 20 mm (0.79 in.) of the sub-base, but no contamination of the underlying soil.

The Water Research Centre at the University of Australia has also carried out research which involved determining the performance of such systems over a long period of time. Their research involved simulating rainfall of 580 mm per annum with a loading of 200 ppm of silts.

The results suggested that after 30 years, the permeability of the surface would only be reduced by 25% (Hart, 2002).

The problem of surface clogging on pervious surfaces is a serious concern. Kobayashi (1999) presented useful details on Japanese experience in the use of manual and automatic washing/suction techniques on both porous asphalt and block surfaces. Kobayashi found that it was possible to raise the infiltration rate from around 0.4 mm/h (0.016 in./hr) to 400 mm/h (16 in./hr), the best performance being achieved with the higher jetting pressures combined with suction (Pratt, 2001).

12.7.4 Maintenance

The maintenance requirements of a pervious surface must be reviewed at the time of design and should be clearly specified. Maintenance is required to prevent clogging of the pervious surface. The factors to be considered when defining maintenance requirements must include the following:

- Type of use;
- Ownership;
- Level of traffic; and
- The local environment and any contributing catchments.

Studies in the United Kingdom have shown satisfactory operation of porous pavement systems without maintenance for over 10 years, and recent work by Imbe et al. at 9th ICUD, Portland, 2002, describes systems operating for more than 20 years without maintenance. However, performance under such regimens could not be guaranteed, and Table 12-6 shows typical recommended maintenance regimens.

12.7.5 Design Life

If maintenance and remedial works are carried out as indicated above, then there is no reason that such systems should not have infinite lifetimes. Current experience suggests that permeable pavements might operate with limited maintenance for 15–20 years, after which time performance would deteriorate rapidly due to clogging.

Table 12-6. Typical Maintenance Activities for Permeable Pavements.

Activity	Schedule
♦ Minimize use of salt or grit for deicing. ♦ Keep landscaped areas well maintained. ♦ Prevent soil being washed onto pavement.	Ongoing
♦ Vacuum clean surface using commercially available sweeping machines at the end of winter (April), midsummer (July/August), and again after autumn leaf fall (November).	2–3 times per year
♦ Inspect catchpits at outlets.	Annually
♦ If routine cleaning does not restore infiltration rates, then reconstruction of part of the whole of a previous surface may be required. ♦ The surface area affected by hydraulic failure should be lifted for inspection of the internal materials to identify the location and extent of the blockage. ♦ Surface materials should be lifted and replaced after brush cleaning. Geotextiles may need complete replacement. ♦ Subsurface layers may need cleaning and replacing. ♦ Removed silts may need to be disposed of as controlled waste.	As needed (infrequent). Maximum 15–20 years.

12.8 Costs

Permeable pavements are up to 25% cheaper (or at least no more expensive than the traditional forms of pavement construction), when all construction and drainage costs are taken into account. (Accepting that the porous macadam itself is a more expensive surfacing, the extra cost of which is offset by the savings in underground pipework, etc.) (Niemczynowicz and Hogland, 1987). Table 12-7 gives U.S. cost estimates for capital and maintenance costs of porous pavements (Landphair et al., 2000).

There is little or no relevant cost information available for the United Kingdom.

12.9 Research Needs

Research needs for permeable pavement are described below.

12.9.1 Design

To assess the long-term stability of pervious pavements in locations more test demonstration sites are required. In particular, the following effects need to be more clearly understood:

- Performance of saturated subgrades;
- Aggregate grading and performance under load;
- Performance of geosynthetic structures;
- Development of a standard set of specifications for materials and installation;
- Effects of polluted runoff: for example, petrol spills on open-graded asphalt or geosynthetic structures;

- Approaches to block replacement; and
- Current design methods are conservative, as account is not taken of vertical and horizontal attenuation processes through the granular subfill, or of the substantial evaporation that appears to occur.

12.9.2 Performance

- Widespread monitoring of the water quality aspects of pervious pavement performance should be a priority to confirm laboratory evidence of quality improvements; assessments should cover a range of different soil types and some lined structures;
- Information lacking also includes seasonal changes in pollutant supply and throughput; and
- The impact of the passage of dissolved pollutants through infiltration structures to soil and groundwater needs to be evaluated.

12.9.3 Construction

- Specific research is needed on geotextiles:
  1. Should there be one at the high level, below the surfacing materials, or is it better to accumulate any sludge at the base of the pavement?
  2. If there is a geotextile, or more than one at different locations, what type and characteristics should be chosen to enhance pollutant-trapping efficiency?
- Groundwater impacts from recently constructed surfaces need to be evaluated.

12.9.4 Maintenance

- The length of the operating life requires further research.
- The influence of operation and maintenance activities on whole-life performance must also be investigated.
- The effectiveness of differing operation and maintenance techniques and equipment should be evaluated.
- The effectiveness of differing maintenance frequencies should be investigated.

12.9.5 Costs

There is limited cost information currently available in the public domain on the installation costs of such systems.

CHAPTER 13.0

BMPs AND DOWNSTREAM ECOLOGY

13.1 Introduction

The adverse impacts of urbanization on biological communities, receiving water quality, and habitat are well known and have been summarized by many authors including Herricks (1995). These changes include higher concentrations of many potentially toxic constituents, degradation of channel bed and bank resulting in habitat loss, and loss of sensitive species. To mitigate this damage, regulatory agencies are now commonly requiring implementation of structural BMPs/SUDS to improve discharge quality and reduce peak flow rates.

One of the major questions associated with implementation of BMPs/SUDS is the degree to which they actually prevent environmental degradation in the receiving water. Many studies have been conducted on BMP performance and substantial removal of pollutants has been observed. However, it is not clear that reduction of concentrations in urban runoff mitigates the biological impact on sensitive species. Horner (1995) notes that even after decades of implementing ambitious protection efforts, resource losses continue.

Even where BMPs remove potential constituents of concern such as sediment, it is not clear that this has a significant impact on in-stream concentrations. For instance, the City of Austin, Texas, requires sand filter or equivalent treatment for new development in most areas of town. The untreated runoff from these areas would typically have about 100 mg/L of TSS, which the sand filter would reduce to approximately 10 mg/L. However, TSS concentrations in creeks in largely undeveloped watersheds have concentrations of about 400 mg/L and creeks in the urban areas often have TSS concentrations that exceed 1500 mg/L (Osborne, Barrett, and Maidment, 2000). These high concentrations demonstrate the significant contribution that bed and bank erosion can make to in-stream water quality and the comparatively small impact that some constituents in surface runoff can have on receiving water conditions.

The purpose of this chapter is to summarize the current state of knowledge regarding the effectiveness of structural BMPs for mitigating ecological impacts of urban runoff in receiving waters.

13.2 Impacts of Urbanization

May and Horner (2002) report that while much research still needs to be done in order to fully understand the linkages between watershed urbanization and ecological integrity, current research in the U.S. Pacific Northwest has identified several key findings, including:

- There is no single cause for the decline of water resource conditions in urbanizing watersheds. Instead, it is the cumulative effects of multiple stressors that are responsible for degraded aquatic habitat and water quality. Imperviousness, while not a perfect yardstick, appears to be a useful predictor of ecological condition. However, a range of stream conditions can be associated with any given level of imperviousness. In general, only streams that retain a significant proportion of their natural vegetative land cover and have very low levels of watershed imperviousness appear to retain their natural ecological integrity. It is this change in watershed land cover that is largely responsible for the shift in hydrologic regime from a subsurface flow dominated system to one dominated by surface runoff.

- There is no clear threshold of urbanization below which there exists a no-effect condition. Instead, there appears to be a relatively continuous decline in almost all measures of water quality or ecological integrity. A low level of watershed imperviousness will not necessarily guarantee a high level of ecological integrity. Undeveloped watersheds with extensive timber harvest, mining, or agricultural activity can also exhibit degraded in-stream habitat, water quality, and/or biological integrity.
- Although the decline in ecological integrity is relatively continuous and is consistent for all parameters, the impact on physical conditions appears to be more pronounced earlier in the urbanization process than chemical degradation. It is generally acknowledged, based on field research and hydrologic modeling, that it is the shift in hydrologic conditions that is the driving force behind physical changes in urban stream-wetland ecosystems.
- There are multiple scales of impact operating within urbanizing watersheds. These include landscape-level impacts, including the loss of natural forest cover and the increase in impervious surface area throughout the watershed. There are also riparian corridor specific impacts, such as encroachment, fragmentation, and loss of native vegetation. In addition, there are also local or backyard impacts such as water diversions, exotic vegetation, stream channelization, streambank hardening (riprap), culvert installation, and pollution from widespread use of pesticides and herbicides. All these stressors contribute to the overall cumulative impact.
- Measures of biological integrity are the most responsive and integrative indicators of overall aquatic ecosystem health. Although much of the environmental concern in the region centers on the condition of native salmonid resources, it is often difficult, due to their transient residence in freshwater habitat, to use these species as direct measures of biological integrity. The use of benthic macroinvertebrates has proven to be an effective surrogate for biological integrity in salmon streams of the U.S. Pacific Northwest.

13.3 Previous Work

Strecker and Urbonas (2001) report that there have been only a limited number of efforts to compile and evaluate quantifiable data about the implementation of BMPs and their impacts on receiving waters. In addition, these past studies did not follow consistent protocols nor did they specifically define the exact causes of the impacts. Strecker and Urbonas report that these studies contributed significantly to the knowledge base for specific study area, but were very limited in scope. These studies are briefly described below.

Maxted and Shaver (1997) were not able to distinguish a biological advantage associated with the presence versus absence of structural BMPs along Delaware streams. However, the design parameters of the BMPs were not presented, so it was not possible to assess how well they controlled surface runoff rates and quality. In contrast, Jones, Via-Norten, and Morgan (1997) concluded that appropriately sited and designed BMPS provided some mitigation of stormwater impacts on Virginia stream habitats and biota, but that the resulting communities were still greatly altered from those in undeveloped reference watersheds.

Because of the uncertainty in the effectiveness of BMPs for preserving biological communities, the EPA commissioned the Watershed Management Institute to investigate stream habitats and biology across gradients of urbanization and BMP application in four regions of the United States (Austin, Texas; Montgomery County, Maryland; Puget Sound, Washington; and Vail, Colorado). In this study biological health was assessed according to the benthic index of biotic integrity and the ratio of salmon to trout, where applicable. Both these measures declined with increases in total impervious area without exhibiting a threshold effect with declines accompanying even small levels of urbanization (Horner et al., 1997; May et al., 1997). An important finding of this study was that until impervious area exceeded 40%, biological decline was more strongly associated with hydrologic fluctuation than with chemical water and sediment quality.

May et al. (2002) compared the benthic index of biological integrity (B-IBI) scores of watersheds in the Puget Sound region of the United States, some with BMPs, some without. The results of this analysis are presented in Figure 13-1 and appear to show little beneficial impact from BMP implementation.

One problem identified by May et al. (2002) was that in many cases only a small portion of runoff from the contributing catchment was treated by the structural BMPs. The data did suggest that more thorough coverage might offer substantive benefits at moderate levels of impervious cover. Their findings also indicate that nonstructural measures such as protection of riparian corridors were not capable of maintaining an ecological system functioning at or very close to the undeveloped conditions, but these measures may help preserve some lower level of health and to prevent degradation to a poor condition. This result echoes the belief of EPA (Swietlik, 2001) that in urban areas with large amounts of imperviousness, loss of riparian cover, extensive habitat modifications, altered hydrology and numerous pollutant and runoff sources, achieving the highest level of ecological integrity may no longer be feasible.

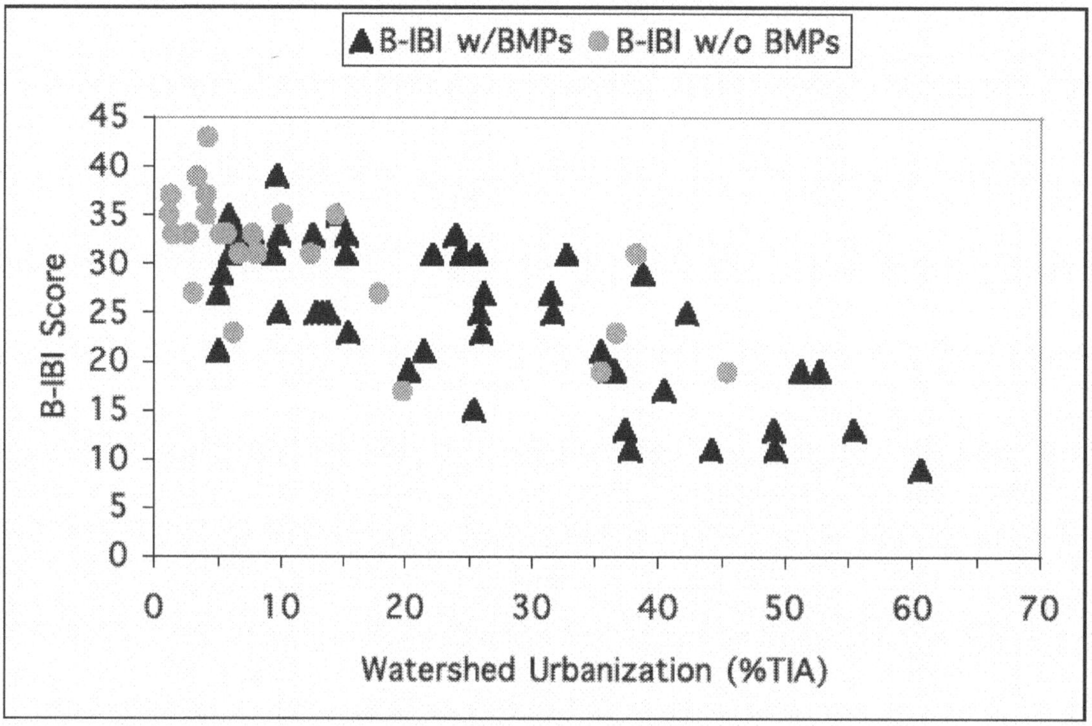

Figure 13-1. Comparison of Biological Integrity Scores (May and Horner, 2002).

Stribling et al. (2001) performed a case study to assess the relationship between BMP implementation and biological condition. They found that it was the combination of the unstable stream channel, and thus habitat, plus the water chemistry problems that were the likely stressors impacting in-stream biology. In contrast to other studies, they were able to demonstrate some improvement in stream biology. Improvement was seen in some selected benthic metrics, and in the fact that several reintroduced fish species seemed to have established reproducing populations.

Mikkelsen et al. (2001) report that there are no empirical studies that document changes of receiving water quality caused solely by stormwater BMPs. They report that the primary factor determining the ecological quality of urban streams is the physical structure of the streams themselves and that the actual effects of BMPs on the receiving water quality are either not well understood or not known at all. They recommend that no further large-scale facilities be constructed until such effects can be determined.

13.4 Ecological Assessment

Burton and Pitt (2001) recently completed a book and associated EPA report compiling guidelines for assessing urban receiving waters affected by urban runoff. Specific experimental designs, plus sampling, laboratory, and statistical tools, and overall assessment methods, comprise most of the content of the book. A summary of the techniques is described by Pitt and Burton (2001). One finding of the study indicates that no one single approach (e.g., chemical-specific criteria, benthic microorganisms, or habitat surveys) can routinely be used to accurately determine or predict ecosystem health and beneficial use impairment. Each assessment approach has associated strengths and weaknesses, so the selection of specific assessment tools and goals is highly dependent on local conditions and objectives.

13.5 Research Needs

Further research on the impact of BMP/SUDS implementation on ecological conditions in the receiving water is a high priority. Current research has documented little benefit to receiving water biological communities from their implementation despite the fact that runoff quality is clearly improved. Since much of the environmental damage appears to result from the change in hydrologic regime and subsequent channel erosion and habitat loss, a key component of any new work in this area should incorporate the effect of BMP capture volume on stream stability.

13.6 Additional Resources

Davis, W., and Simon, T., 2001, *Biological Assessment and Criteria: Tools for Water Resource Planning and Decision Making*, Lewis Publishers, Boca Raton, FL.

Horner, R., May, C., Livingston, E. Blaha, D., Scoggins, M. Tims, J. and Maxted, J., 2001, Structural and Non-Structural BMPs for Protecting Streams. In Proceedings of Engineering Foundation Conference: Linking Stormwater BMP Designs and Performance to Receiving Water Impact Mitigation, August 19-24, Snowmass Village, CO, pp. 60-77.

Horner, R., and May, C., 1999, Regional Study Supports Natural Land Cover Protection as Leading Best Management for Maintaining Stream Ecological Integrity. In Proceedings of the Comprehensive Stormwater and Aquatic Ecosystem Management Conference, Auckland, New Zealand.

APPENDIX B

SUDS PLANNING AND APPROVAL IN THE UNITED KINGDOM

A drainage project in England and Wales can involve a number of different bodies, such as private landowners, local authorities, sewerage undertakers, and the Environment Agency. Environmental and conservation responsibilities are defined in national and international legislation; Agenda 21 promotes them at a local level. There is, however, no specific legislation pertaining to drainage sustainability, since the current drainage legislation was enacted before the use of SUDS became widespread. While this lack of legislation poses no legal difficulty, the responsibilities for provision, operation, and maintenance of SUDS are not clearly defined. Successful designs involve a partnership of members of the development team; adoption and maintenance of a system require a similar joint approach.

1.1 Responsibilities

Public drainage responsibilities are divided among four main bodies:

- Local authorities
- Highways authorities
- Sewerage utilities
- The Environment Agency

Local authorities carry out some or all of the following functions (depending upon whether they work within a two-tier or a single-tier system): planning (policy and development control), highway authority, the management of leisure and landscape land, environmental health and land drainage ('nonmain river'). The Highways Agency controls drainage of trunk roads, while sewerage utilities are responsible for the sewers that convey surface water from impermeable areas, such as roofs and drives, on private property. New development is controlled by local planning departments, with allowable discharges and consents negotiated with the Environment Agency or sewerage utility. The Environment Agency is responsible for flood protection and land drainage, including supervision over all flooding, but primarily with respect to main rivers.

Local authorities are in charge of implementing Agenda 21 and development of strategies to secure sustainability at a local level. This principle applies to drainage, and sustainable urban drainage is increasingly being made a part of Development Plans and regional planning guidance.

1.2 Ownership and Maintenance

The ownership and maintenance of conventional piped drainage systems is clearly defined in Sewers for Adoption (Water Services Association, 1994). However, many SUDS can be defined as either drainage or landscape features, which makes it difficult to determine who is responsible for their operation and maintenance. In Scotland, a trial framework agreement has been outlined to remove this barrier to the greater use of SUDS.

Because of the numerous legal issues involved, researching a countrywide agreement (pertaining to ownership and maintenance) in England and Wales will take time; however, individual maintenance agreements can be negotiated on a site-by-site basis.

1.3 Approvals

Formal approval of a drainage system is required from the following authorities:

- The planning authority (planning permission)
- Building control authority (approval of design and implementation of the Building Regulations)
- The sewerage undertaker (connection to the adopted sewers and possible adoption of drainage systems)
- The highway authority (permits for road construction and for highway drainage)
- Local authority (possible agreement to maintain public open space)
- The Environment Agency (discharge consents may be needed in some cases)

The planning process may be used by other interested parties, such as nongovernmental organizations and pressure groups to promote their views.

The planning system is used to coordinate consultation between the approving authorities, but the license and the consents must be applied for separately; they are not granted automatically when a planning application is approved.

1.4 Planning Authority

The planning authority defines local and structure plans, and identifies areas for development and the standards for such development in terms of land use and building density. It also controls development through the process of application, negotiation, and approval (with or without conditions/agreements) or refusal. These standards are based on judgments rather than strict regulations. The clear plans allow developers to make decisions regarding site development. Specification of SUDS in a plan will enable the developer to assess the impact of the drainage system at an early stage, which may be important in terms of the land requirement of the drainage devices.

Planning authorities can set design criteria for the drainage system that will fulfill some amenity objectives, such as public open space. In appropriate circumstances, a planning authority may seek agreements to ensure continued maintenance. The drainage criteria should fit into an overall strategy. Planning Policy Guidance (PPG) 25 Development and flood risk (2001) advises developers to assess the implications of runoff from their proposed development and to control it through sustainable drainage systems where possible. The PPG also advises that local plans should include policies that promote SUDS.

1.5 Building Control

Before construction can proceed, building control bodies must be satisfied that 1) adequate provisions have been made for drainage, and 2) the proposed system will not affect the integrity of buildings and meets the requirements of The Building Regulations 1991 (as amended) (The Stationery Office). Building control departments will, therefore, be interested in the location of drainage systems and their proximity to structures; thus, the proximity of infiltration devices may be important in unstable soils. Interpretation of The Building Regulations might vary among authorities. Forthcoming amendments to Part H of the Building Regulations will give greater priority to the implementation of SUDS. Part H now sets out a hierarchy for surface water drainage within SUDS as the first choice.

1.6 Highway Authority

If a development includes road drainage, a consent should be obtained from the highway authority (usually a department of the local government). Trunk roads are the responsibility of the Highways Agency, which supports SUDS-type facilities and promotes their use through its Design Manual for Roads and Bridges, but prefers not to refer to the facilities as "SUDS."

Each authority establishes standards that developers must follow throughout construction to ensure that adoptable roads are of satisfactory construction, safe for public use, and easily maintained. Effective road drainage is essential to the safety as well as the integrity and structural stability of the road, including its footways, verges, and margins. Local authorities, when considering applications for construction permits, need to be satisfied that the SUDS in each location meet their road drainage requirements and will not require extensive maintenance.

1.7 Local Authority

The Local Authorities are involved in flood protection, land drainage, and maintenance of public open space.

1.8 Sewerage Undertaker

The sewerage undertaker (usually the water service company) approves plans of those components of the drainage system that will be adopted under Section 104 of the Water Industry Act, and negotiates connection to or adoption of sewers that are covered by the document, Sewers for Adoption (Water Services Association, 1994).

1.9 Environment Agency

The Environment Agency is responsible for the protection of controlled waters in England and Wales, and will be a consultant to the Planning Authority on drainage issues. It can provide general advice on pollution potential within a catchment and on water pollution prevention practices. Some discharges may be subject to a Water Resources Act 1991 permit or a Groundwater Regulations 1998 authorization from the Agency. The agency also has the authority to serve notices requiring implementation of measures to protect water courses and groundwater quality.

1.10 Other Statutory Environmental Organizations

English Nature and the Countryside Council for Wales/Cyngor Cefn Gwlad Cymru are the governments' statutory advisers on sustaining natural beauty, wildlife, and the opportunities for outdoor enjoyment. They can offer guidance on the amenity value of proposed developments.

Runoff from industrial areas may be subject to a prohibition notice if the water remains contaminated, especially if there is a likelihood of groundwater pollution. Flood protection permits may also be required to regulate runoff to certain watercourses. Any permit will likely be of a descriptive rather than a numeric nature. The Environment Agency should be consulted regarding all discharges to marine waters, watercourses, and aquifers. In some cases, where surface water is likely to be contaminated, an authorization may be required.

2.1 Planning and Approval—Scotland and Northern Ireland

In Scotland, the responsibility for drainage is shared by several authorities. Recent discussions with these authorities have clarified the maintenance responsibilities for new drainage systems (in most cases) for a trial period. The key to success is consultation among developers, professionals, and regulatory authorities. Consultation produces consensus on design criteria and leads to the selection of appropriate drainage techniques for each development. In Northern Ireland, although different authorities carry these responsibilities and different legislation applies, consultation is still necessary. Scottish procedures can be used as a guide.

2.2 Responsibilities

Since local government reorganization in April 1996, surface water drainage responsibilities outside private properties are divided among the following three main bodies:

- Local authorities
- Scottish Water
- The Scottish Environment Protection Agency, SEPA

Local authorities act as planning, building control, and roads authorities. New development, including drainage, requires the approval of local authority planning departments. Roads authorities, Scottish Water, and SEPA are statutory consultants to the planning process. The planning process, therefore, is used to coordinate drainage control for new developments.

Local authority-led Flood Appraisal Groups coordinate action where catchments include several separate local authorities. NPPG7 (policy guidance on flooding and planning) identifies some of the planning and flood management issues.

Some authorities are committed to implementing SUDS programs and are working together to promote their use. SUDS are now featured in some local plans and in planning authorities' policy guidance. SUDS will be eventually incorporated into local plans, which will reduce the need for early consultation, as the regulatory authorities' criteria for drainage design will have already been defined. In the meantime, regulatory authorities are forging agreements and improving the approval process for developers.

2.3 Ownership and Maintenance

Discussions between authorities have been instrumental in solving problems caused by drainage constraints. For example, some local authorities and Scottish Water have agreements for shared drainage that allow a single system to drain water from private properties and roads. Under these agreements, shared surface water sewers are owned and maintained by Scottish Water. The local authority oversees pipes that serve only road drains and empty the gully pots.

The most helpful drainage legislation is Section 7 of the Sewerage (Scotland) Act, which requires the roads agency and Scottish Water to enter into agreements for shared drainage and is the basis for public SUDS maintenance agreement.

Ownership and maintenance responsibilities for systems that are not covered by shared drainage agreements are as follows:

- Owners are responsible for drainage within the curtilage of their properties
- Roads authorities are responsible for drainage of adopted roads
- Water authorities are responsible for drainage beyond the curtilage of properties (unless the site is served by a private sewer)

Ownership of the land within the curtilage of the property may also change through the development process, with a landowner selling the land to a developer, who then sells it to the final owner. Each party should be aware of any drainage, land use, or maintenance.

2.4 Approvals

Formal approval for a drainage system is needed from the following institutions:

- The planning authority (planning permission)
- Building control authority (building warrant)
- Scottish Water (drainage construction consent)
- The roads authority (road construction consent)
- SEPA (consent to discharge, if required, or conditional prohibition notice)

Each regulatory authority has its own powers; there is no hierarchy of statutory instruments.

2.5 Planning Authority

The planning authority establishes structure and local plans and identifies areas for different land uses in local plans.

The planning system is used to coordinate consultation among the approving authorities, but the license and consents require separate application; they are not granted automatically upon approval of the planning. Other interested parties, such as members of the public and nongovernmental organizations, also utilize the planning process to promote their agendas.

Planning authorities can set criteria for the design of the drainage system that will fulfill some of their amenity objectives, such as public open space. In certain circumstances, a planning authority may seek agreements to ensure continued maintenance.

The Scottish Executive has published Planning Advice Note (PAN) 61 Planning and Sustainable Urban Drainage Systems (July 2001), which contains advice for both planners and the development industry.

2.6 Building Control Authority

Building control officers require adequate provisions for drainage. The system should be designed and constructed to meet the Technical Standards for compliance with the Building Standards (Scotland) Regulations 2002 (as amended). SUDS for areas with more than 200 m^2 of "hard surface" are specified in Part M. Building control departments (building quality departments in some authorities) will, therefore, be interested in the location of drainage systems and their proximity to structures. Interpretation of the Building Regulations might vary between authorities.

2.7 Scottish Water

Scottish Water gives technical approval of the proposed sewerage infrastructure to serve the development. The infrastructure will generally be accepted, provided that it is constructed in accordance with the approved layout and with Scottish Water's specification.

2.8 Roads Authority

Construction consent from the local authority (such as the roads authority) is required to construct or alter a road that will be adopted by the authority. Connection of a private road or access to the local road network will require the authority's permission. Trunk roads are controlled by the Scottish Executive (through the Roads Agency in Scotland), whose permission to alter or to connect to a trunk road should be sought. As with the Highways Agency in England and Wales, the Roads Agency promotes the implementation and use of SUDS-type facilities through its Design Manual for Roads and Bridges, although it prefers not to refer to the facilities as "SUDS."

Each authority establishes standards that developers must follow throughout construction to ensure that the roads are of satisfactory construction, safe for public use, and easily maintained. Effective drainage is fundamental to the safety, integrity, and structural stability of the road, including its footways, verges, and margins. When considering applications for construction permits, local authorities require that the SUDS in the locations under their jurisdiction meet their road drainage requirements and not require extensive maintenance.

2.9 SEPA

The Control of Pollution Act 1974 (as amended) gives SEPA powers to permit discharges to controlled waters. The Act defines these three categories of discharge:

- Trade effluent
- Sewage effluent
- Matter other than trade or sewage effluent. Surface water discharges are regarded as "other matter"

Discharges of "other matter" to controlled waters do not require a permit unless SEPA serves a prohibition notice. Such a notice may be conditional or absolute (requiring a discharger to apply for consent). Discharges of "other matter" to land are not subject to SEPA's statutory control.

Even if a consent to discharge is unlikely to be required, SEPA should be made aware of the drainage proposals. SEPA's policy is to require SUDS to be considered at the planning stage of any proposed development.

2.10 Other Statutory Environmental Organizations

Scottish Natural Heritage is the Scottish Executive's statutory advisor on sustaining natural beauty, wildlife, and the opportunity for outdoor enjoyment. They promote SUDS issues and can offer advice on the amenity value of proposed developments.

REFERENCES

Abbott, C.L., Comino-Mateos, L., 2001, In situ performance monitoring of an infiltration drainage system and field testing of current design procedures. *Journal CIWEM*, 15(3), pp.198-202.

Abbott C.L., Comino L., Angood C., 2000, *Monitoring Performance of Infiltration Drainage Systems,* Report SR569HR. Wallingford, U.K.

Allgemeine Technische Vorschriften (ATV), 2001, *Report 138 Planung, Bau und Betrieb von Anlagen zur Versickerung von Niederschlagswasser,* Gesellschaft zur Foerderung der Abwassertechnik e.V., Hennef, Germany.

American Petroleum Institute (API), 1990, *Monographs on Refinery Environmental Control—Management of Water Discharges (Design and Operation of Oil-Water Separators).* Publication 421, First Edition.

American Society of Civil Engineers (ASCE), 2002, *Urban Stormwater BMP Performance Monitoring: A Guidance Manual for Meeting the National Stormwater BMP Database Requirements*. EPA-821-B-02-001. Prepared by GeoSyntec Consultants, Urban Drainage and Flood Control District, and Urban Water Resources Research Council (UWRRC) of ASCE in cooperation with U.S. EPA, Office of Water, Washington, D.C.

AquaLogic Stormwater Filtration System Engineering Manual, 2000, AquaLogic Stormwater Abatement Filter Systems. San Antonio, TX.

Aronson, G., Watson, D., Pisaro, W. *Evaluation of Catch Basin Performance for Urban Stormwater Pollution Control*. U.S. EPA, Washington, D.C.

Backstrom, M., 2001, Particulate trapping in grass swales. *Novatech 2001-Innovative technologies in urban drainage conference proceedings, 25-27 June 2001, Lyon, France.* GRAIE, Villeurbanne, Lyon, France. ISBN 2 9509337-4-2. pp. 391-398.

Balades, J.D., Petitnicolas, F., 2001, *Strategies of wet weather pollution control at source. A technical and financial approach. Novatech 2001-Innovative technologies in urban drainage conference proceedings, 25-27 June 2001, Lyon, France.* GRAIE, Villeurbanne, Lyon, France. ISBN 2 9509337-4-2. pp. 367-373.

Bannerman, R., Dodds, R.,1992, Unpublished data. Bureau of Water Resources Management, Wisconsin Department of Natural Resources, Madison, WI.

Baptista, M., Barraud, S., Alfakih, E., 2001, Data analysis for the elaboration of technical and economic indicators of alternative storm drainage systems. *Novatech 2001-Innovative technologies in urban drainage conference proceedings, 25-27 June 2001, Lyon, France.* GRAIE, Villeurbanne, Lyon, France. ISBN 2 9509337-4-2. pp 63.70.

Barrett, M., 1999, *Complying with the Edwards Aquifer Rules: Technical Guidance on Best Management Practices*, Texas Natural Resource Conservation Commission Report RG-348. http://www.tnrcc.state.tx.us/admin/topdoc/rg/348/index.html.

Barrett, M.E., Walsh, P.M., Malina, J.F., Jr., Charbeneau, R.J., 1998, Performance of vegetative controls for treating highway runoff, *ASCE Journal of Environmental Engineering*, Vol. 124, No. 11, pp. 1121-1128.

Barton Springs/Edwards Aquifer Conservation District, 1996, *Final Report: Enhanced Roadway Runoff Best Management Practices*. City of Austin, Drainage Utility, LCRA, TDOT. Austin, TX.

Bell, W., L. Stokes, L.J. Gavan, T.N. Nguyen, 1995, *Assessment of the Pollutant Removal Efficiencies of Delaware Sand Filter BMPs*. Final Report. Department of Transportation and Environmental Services. Alexandria, VA. 140 pp. Also in Performance of Delaware Sand Filter Assessed. *Watershed Protection Techniques*. Center for Watershed Protection. Fall 1995. Vol. 2(1): 291-293.

Berg, V.H, 1991, *Water Quality Inlets (Oil/Grit Separators)*. Maryland Department of the Environment, Sediment and Stormwater Administration.

Borden, R.C., Dorn, J.L., Stillman, J.B., Liehr, S.K., 1996, *Evaluation of Ponds and Wetlands for Protection of Public Water Supplies.* Draft Report. Water Resources Research Institute of the University of North Carolina, Department of Civil Engineering, North Carolina State University, Raleigh, NC.

Brown, W., Schueler, T., 1997, *The Economics of Stormwater BMPs in the Mid-Atlantic Region*. Prepared for the Chesapeake Research Consortium, Edgewater, MD, by the Center for Watershed Protection, Ellicott City, MD.

Building Research Establishment (BRE), 1991, *Digest 365 Soakaway Design.*

Burton, G., Pitt, R., 2001, *Stormwater Effects Handbook: A Toolbox for Watershed Managers, Scientists, and Engineers*, CRC Press, Boca Raton, FL, 985 pp.

Butler, D., Davies, J.W., 2000, *Urban Drainage.* E & FN Spon. London.

Butler, D., Memon, F., 1999, Dynamic Modelling Of Roadside Gully Pots During Wet Weather, *Water Research* Vol. 33, No. 15, pp. 3364-3372, Elsevier Science Ltd.

Caltrans, 2002, *BMP Retrofit Pilot Program Proposed Final Report*, Report CTSW-RT-01-050 California Dept. of Transportation, Sacramento, CA.

Caltrans, 2001, *Third Party Best Management Practice Retrofit Pilot Study Cost Review*, Appendix to BMP Retrofit Proposed Final Report, California Dept. of Transportation, Sacramento, CA.

Center for Watershed Protection (CWP), 1998, *Costs and Benefits of Storm Water BMPs*. Ellicott City, MD.

Center for Watershed Protection (CWP), 1997, Multi-Chamber Treatment Train developed for stormwater hot spots. *Watershed Protection Techniques* 2(3): 445-449.

Center for Watershed Protection (CWP), 1996, *Design of Stormwater Filtering Systems*. Prepared for the Chesapeake Research Consortium, Solomons, MD, and U.S. EPA Region V, Chicago, IL, by the Center for Watershed Protection, Ellicott City, MD.

Chocat, B., Barraud, S., Alfakih, E., 2001, Development of BMPs: Progress in France and Western Europe, in proceedings of an Engineering Foundation Conference: Linking Stormwater BMP Designs and Performance to Receiving Water Impact Mitigation, August 19-24, Snowmass Village, CO.

City of Austin, Texas, 1996a, Evaluation of Non-Point Source Controls: A 319 Grant Project. Draft Water Quality Report Series, Public Works Department, Austin, TX.

City of Austin, Texas, 1996b, *Design of Water Quality Controls*. City of Austin, TX.

City of Austin, Texas, 1991, *Design Guidelines for Water Quality Control Basins*. Public Works Department, Austin, TX.

City of Austin, Texas, 1990, *Removal Efficiencies of Stormwater Control Structures*. Final Report. Environmental Resource Management Division. 36 p. Also in: Developments in Sand Filter Technology to Improve Stormwater Runoff Quality. *Watershed Protection Techniques*. Center for Watershed Protection. Summer 1994. Vol. 1(2): 47-54.

Clark, S.E., 2000, Urban Stormwater Filtration: Optimization of Design Parameters and a Pilot-Scale Evaluation, Ph.D. Dissertation, University of Alabama at Birmingham.

Codling A., 1992, Design and construction of infiltration basins at A35 Yellowhan Hill. Proceedings of the Standing Conference on Stormwater Source Control, Volume V, Coventry University.

Coffman, L.S., Goo, R., Frederick, R., 1999, Low impact development: an innovative alternative approach to stormwater management. Proceedings of the 26th Annual Water Resources Planning and Management Conference ASCE, June 6-9, Tempe, AZ.

Colwell, S.R., Horner, R.R., Booth, D.B., 2000, *Characterization of Performance Predictors and Evaluation of Mowing Practices in Biofiltration Swales*. Report to King County Land and Water Resources Division and others by Center for Urban Water Resources Management, Department of Civil and Environmental Engineering, University of Washington, Seattle, WA.

Construction Industry Research and Information Association (CIRIA), 2002, *Source Control using Constructed Pervious Surfaces, Report C582,* London, SW1P 3AU.

Construction Industry Research and Information Association (CIRIA), 2000a, *Sustainable urban drainage systems—design manual for Scotland and Northern Ireland, Report C521,* London, SW1P 3AU.

Construction Industry Research and Information Association (CIRIA), 2000b, *Sustainable urban drainage systems—design manual for England and Wales, Report C522,* London, SW1P 3AU.

Construction Industry Research and Information Association (CIRIA), 1999, *Control of pollution from highway drainage discharge, Report 142,* London, SW1P 3AU.

Construction Industry Research and Information Association (CIRIA), 1996, *Infiltration Drainage: Manual of Good Practice, London,* SW1P 3AU.

Construction Industry Research and Information Association (CIRIA), 1995, *Sediment management in urban drainage catchments, Report 134,* London, SW1P 3AU.

Construction Industry Research and Information Association (CIRIA), 1993, *Design of Floor Storage Reservoirs, Book 14,* London, SW1P 3AU.

Construction Industry Research and Information Association (CIRIA), no year, *Manual of good practice for the design, construction, and maintenance of infiltration drainage systems for stormwater runoff control and disposal, Report 448,* London, SW1P 3AU.

CSF Treatment Systems, Inc. (CSF), 1996, *Stormwater management promotional brochure.* CSF Treatment Systems, Inc., Portland, OR.

Cullum, M., 1985, *Stormwater Runoff Analysis at a Single Family Residential Site.* Publication 85-1. University of Central Florida, Orlando, FL. pp. 247–256.

Curran, T., 1996, Peat Sand Efficiency Calculations for McGregor Park. Unpublished data. Lower Colorado River Authority. Austin, TX.

D'Arcy, B.J., Frost, C.A., 2001, The Role of Best Management Practices in Alleviating Water Quality Problems Associated with Diffuse Pollution, *Science of the Total Environment* 265, pp. 359-367.

Davis, A.P., Shokouhian, M., Sharm, H., Minami, C., 2001, Laboratory study of biological retention for urban stormwater management. Water Environment Research, 73(1), 5-14.

Davis, A.P., Shokouhian, M., Sharma, H., Minani, C., 1998, *Optimization of Bioretention Design for Water Quality and Hydrologic Characteristics.* Final Report. Environmental Engineering Program, Department of Civil Engineering, University of Maryland, College Park, Maryland.

Davis, W., Simon, T., 2001, *Biological Assessment and Criteria: Tools for Water Resource Planning and Decision Making*, Lewis Publishers, Boca Raton, FL.

Denver Urban Drainage and Flood Control District, 1992, *Urban Storm Drainage Criteria Manual—Volume 3: Best Management Practices*. Denver, CO.

Desbonette, A., Pogue, P., Lee, V., Wolff, N., 1994, *Vegetated Buffers in the Coastal Zone: A Summary Review and Bibliography*. Coastal Resources Center. University of Rhode Island, Kingston, RI.

Dierkes, C., Kuhlmann, L., Kandasamy J., Angelis, G., 2002, Pollution Retention Capability and Maintenance of Permeable Pavements. *Proc 9th International Conference on Urban Drainage*, Portland, OR.

Dorman, M.E., Hartigan, J., Steg, R.F., Quasebarth, T., 1996, *Retention, Detention and Overland Flow for Pollutant Removal From Highway Stormwater Runoff. Vol. 1.* Research Report. FHWA/RD 96-095. Federal Highway Administration, Washington, D.C.

Dorman, M.E., Hartigan, J., Steg, R.F., Quasebarth, T., 1989, *Retention, Detention and Overland Flow for Pollutant Removal From Highway Stormwater Runoff. Vol. 1.* FHWA/RD 89/202. Federal Highway Administration, Washington, D.C.

Driscoll, E.D., 1983, *Performance of Detention Basins for Control of Urban Runoff Quality.* Presented at the 1983 International Symposium on Urban Hydrology, Hydraulics and Sedimentation Control, University of Kentucky, Lexington, KY.

Duchene, M., McBean, E., Thomson, N.R., 1994, Modelling of Infiltration from infiltration trenches for stormwater control. *Journal of Water Resources Planning and Management* 120(3): 276-293.

Duchene, M., McBean, E., 1992, Discharge characteristics of perforated pipe for use in infiltration trenches. *Water Resource Bulletin* 28(3): 517-524.

Ellis, J.B., Shutes, R.B., Revitt, D.M., Foreshaw, M., Winter, B., 2001, SUDS and constructed wetlands: how compatible are they? First National SUDS Conference. Coventry University, U.K.

Emmerling-Dinovo, C., 1995, Stormwater detention basins and residential locational decisions. *Water Resources Bulletin*, 31(3):515-52.

Federal Highway Administration (FWHA), 1999, *Environmental Technology Brief.* U.S. Department of Transportation, Federal Highway Administration, Washington, DC.

Folkes-Skinner, A.B., 2002, Environment Agency Monitoring 2000-2002, London.

Gain, W.S., 1996, *The Effects of Flow Path Modification on Water Quality Constituent Retention in an Urban Stormwater Detention Pond and Wetland System.* Water Resources Investigations Report 95-4297. U.S. Geological Survey, Tallahassee, FL.

Galli, J., 1992, *Analysis of Urban BMP Performance and Longevity in Prince George's County, Maryland.* Metropolitan Washington Council of Governments, Washington, D.C.

Galli, F., 1990a, *Thermal Impacts Associated with Urbanization and Stormwater Best Management Practices*. Prepared for the Maryland Department of the Environment, Baltimore, MD, by the Metropolitan Council of Governments, Washington, D.C.

Galli, F., 1990b, *Peat-Sand Filters: A Proposed Stormwater Management Practice for Urban Areas*. Metropolitan Washington Council of Governments, Washington, D.C.

Glick, R., 2001, personal communication, City of Austin Watershed Protection Dept., Austin, TX.

Glick, R., Chang, G.C., Barrett, M.E., 1998, Monitoring and evaluation of stormwater quality control basins, in *Watershed Management: Moving from Theory to Implementation*, Denver, CO, May 3-6, 1998, pp. 369-376.

Goldberg J.S., 1993, *Dayton Avenue Swale Biofiltration Study*. Seattle Engineering Department, Seattle, WA.

Greb, S., Corsi, S., Waschbush, R. 1998. Evaluation of Stormceptor and Multi-Chamber Treatment Train as Urban Retrofit Strategies. Presented at Retrofit Opportunities for Water Resource Protection in Urban Environments, A National Conference. The Westin Hotel, Chicago, IL, February 10-12, 1998.

Hall, M.J., Hockin, D.L., Ellis, J.B., 1993, *Design of Flood Storage Reservoirs.* CIRIA. Book 14. East Kilbride, Scotland. ISBN 0 7506 1057 3.

Haneda M., Kagawa, A., Maruyama, K., 1996, Study on durability of underground infiltration facilities through long-term survey. Proc. 7th Int. Conf. on Urban Storm Drainage, Hannover, Germany, 599-604.

Harper, H., 1988, *Effects of Stormwater Management Systems on Groundwater Quality.* Prepared for Florida Department of Environmental Regulation, Tallahassee, FL, by Environmental Research and Design, Inc., Orlando, FL.

Harper, H., Herr, J., 1993, *Treatment Efficiency of Detention With Filtration Systems.* Environmental Research and Design, Inc. Final Report Submitted to Florida Department of Environmental Regulation. Orlando, FL. 164 pp.

Harper, H., Yousef, Y., Wanielista, M., 1984, Efficiency of Roadside Swales in Removing Heavy Metals from Highway Associated Nonpoint Source Runoff. AWRA Options for Reaching Water Quality Goals.

Hart, P., 2002, Permeable Paving as a Stormwater Source Control System. Paper presented at Scottish Hydraulics Study Group 14th Annual seminar, SUDS. 22 March 2002, Glasgow.

Herricks, E., 1995, *Stormwater Runoff and Receiving Systems: Impact, Monitoring, and Assessment*, CRC Press, Boca Raton, FL.

Hightshoe, G.L., 1988, *Native Trees, Shrubs, and Vines for Urban and Rural America.* Van Nostrand Reinhold, New York.

Hilding, K., 1996, Longevity of infiltration basins assessed in Puget Sound. *Watershed Protection Techniques* 1(3):124-125.

Holler, J.D., 1990, Nonpoint source phosphorus control by a combination wet detention/filtration facility in Kissimmee, FL. *Florida Scientist* 53(1):28-37.

Holler, J.D., 1989, Water quality efficiency of an urban commercial wet detention stormwater management system at Boynton Beach Mall in South Palm Beach County, FL. *Florida Scientist* 52(1):48-57.

Holmstrand O., 1984, Infiltration of stormwater: research at Chalmers University of Technology, results and examples of application. In P. Balmer, P-A Malmqvist, and A. Sjoberg (Editors), *Planning and Control of Urban Storm Drainage,* 1057-1066.

Honeyman, M.J., Baird, I.A., 2001, Newhall Valley Water Management Strategy. First National Conference on Sustainable Drainage, Coventry, U.K.

Horner, R.R., 1995, Toward Ecologically Based Urban Runoff Management. In *Stormwater Runoff and Receiving Systems: Impact, Monitoring, and Assessment,* E.E. Herricks (Editor), CRC Press, Boca Raton.

Horner, R., May, C., Livingston, E. Blaha, D., Scoggins, M. Tims, J., Maxted, J., 2001, Structural and Non-Structural BMPs for Protecting Streams, in proceedings of an Engineering Foundation Conference: Linking Stormwater BMP Designs and Performance to Receiving Water Impact Mitigation, August 19-24, Snowmass Village, CO, pp. 60-77.

Horner, R.R., Horner, C.R., 1999, Performance of a Perimeter (Delaware) Sand Filter in Treating Stormwater Runoff from a Barge Loading Terminal. *Proc. of the Comprehensive Stormwater and Aquatic Ecosystem Management Conf.*, Auckland, N.Z., Feb. 1999, pp. 183-192.

Horner, R., May, C., 1999, Regional Study Supports Natural Land Cover Protection as Leading Best Management for Maintaining Stream Ecological Integrity, *Proceedings of the*

Comprehensive Stormwater and Aquatic Ecosystem Management Conference, Auckland, New Zealand.

Horner, R., Booth, D., Azous, A., May, C., 1997, Watershed Determinants of Ecosystem Functioning, in L. A. Roesner (ed.), *Effects of Watershed Development and Management of Aquatic Ecosystems,* ASCE, NY, pp. 251-274.

Horner, R.R., Horner, C.R., 1995, *Design, Construction and Evaluation of a Sand Filter Stormwater Treatment System.* Part II. *Performance Monitoring.* Report to Alaska Marine Lines, Seattle, WA. 38 p. Also in Performance of Delaware Sand Filter Assessed. *Watershed Protection Techniques.* Center for Watershed Protection. Fall 1995. Vol. 2(1): 291-293.

Horner, R.R., Skupien, J.J., Livingston, E.H., Shaver, H.E., 1994, *Fundamentals of Urban Runoff management: Technical and Institutional Issues.* Terrene Institute, Washington, D.C.

Horner, R.R., Guedry, J., Kortenhoff, M.H., 1990, *Improving the Cost Effectiveness of Highway Construction Site Erosion and Pollution Control.* Final Report. Washington State Transportation Commission, Olympia, WA.

Jefferies, C. (Editor), 2001, *SUDS Monitoring Programme.* SNIFFER Report (00)10. Section 3.3 Detention Basins; Section 5.3 Performance Observations—Detention Basins. Edinburgh EH3 6RT.

Jones, R., Via-Norten, A., Morgan, D., 1997, Bioassessment of BMP Effectiveness in Mitigating Stormwater Impacts on Aquatic Biota. In L.A. Roesner (Editor), *Effects of Watershed Development and Management of Aquatic Ecosystems,* American Society of Civil Engineers, New York, pp. 494-512.

Kaighn, R.J., Jr., Yu, S.L., 1996, Testing of roadside vegetation for highway runoff pollutant removal. *Trans. Res. Rec.*, 1523, 116-123.

Kantrowitz, I., Woodham, W., 1995, *Efficiency of a Stormwater Detention Pond in Reducing Loads of Chemical and Physical Constituents in Urban Streamflow, Pinellas County, Florida.* Water Resources Investigations Report 94-4217. U.S. Geological Survey, Tallahassee, FL.

Keblin, M.V., Barrett, M.E., Malina, J.F., Jr., Charbeneau, R.J., 1998, *The Effectiveness of Permanent Highway Runoff Controls: Sedimentation/Filtration Systems*, Research Report 2954-1, Center for Transportation Research, University of Texas at Austin.

Kercher, W.C., Landon, J.C., Massarelli, R., 1983, Grassy swales prove cost-effective for water pollution control. *Public Works*, 16: 53-55.

King County, Washington, Department of Natural Resources, 2000, *King County Surface Water Design Manual.* [http://splash.metrokc.gov/wlr/dss/manual.htm]. Last updated March 6, 2000. Accessed January 5, 2001.

Kobayashi, M., 1999, Stormwater runoff control in Nagoya City. *Proc. 8th Int. Conf. on Urban Storm Drainage,* Sydney, Australia, pp.825-833.

Koon, J., 1995, *Evaluation of Water Quality Ponds and Swales in the Issaquah/East Lake Sammamish Basins.* King County Surface Water Management, Seattle, WA, and Washington Department of Ecology, Olympia, WA.

Lager, J., Smith, W., Finn, R., Finnemore, E., 1977, *Urban Stormwater Management and Technology: Update and Users' Guide.* Prepared for U.S. Environmental Protection Agency. EPA-600/8-77-014. 313 pp.

Landphair, H., McFalls, J., Thompson, D., 2000, *Design Methods, Selection, and Cost Effectiveness of Stormwater Quality Structures,* Texas Transportation Institute Research Report 1837-1, College Station, TX.

Lawrence, A.J., Marsalek, J., Ellis, J.B., Urbonas, B., 1996, Stormwater Detention and BMPs. *J.Hydr.Res. Vol. 34, No. 6 pp 799-813.*

Legret, M., and Colandini, V., 1998, *Effects of a Porous Pavement with Reservoir Structure on Runoff Water: Water Quality and the Fate of Heavy Metals.* Laboratoire Central Des Ponts et Chaussees.

Leif, T., 1999, *Compost Stormwater Filter Evaluation.* Snohomish County, Washington, Department of Public Works, Everett, WA.

Low Impact Development (LID) Center, 2003, Low impact development (LID) urban design tools (revised). [http://www.lid-stormwater.net/]. Beltsville, MD.

Macdonald, K., 2002, Doctoral thesis, in preparation. University of Abertay, Dundee, Scotland.

Macdonald, K., Jefferies, C., 2002, Performance of Two Swales. Scottish Hydraulics Study Group, 14th Annual Seminar.

Macdonald, K., Jefferies, C., 2001, Performance Comparison of Porous Paved and Traditional Car Parks. *Proc. First National Conference on Sustainable Drainage Systems,* Coventry June 2001.

Macdonald K., Jefferies, C., 2000, The effectiveness of sustainable urban drainage systems in Scotland—Results of flowrate and water quality monitoring at four SUDS sites in Scotland. Fifteenth European Junior Scientist Workshop "Decision Support for Urban Water Management". May 11-14, Terschelling, Netherlands.

MacRae, C., 1996, Experience from Morphological Research on Canadian Streams: Is Control of the Two-Year Frequency Runoff Event the Best Basis for Stream Channel Protection? In L. Roesner (ed.), *Effects of Watershed Development and Management on Aquatic Ecosystems.* American Society of Civil Engineers. Snowbird, UT. pp. 144-162.

Magette, W., Brinsfield, R., Palmer, R., Wood, J., 1989, Nutrient and sediment removal by vegetated filter strips. *Transactions of the American Society of Agricultural Engineers* 32(2): 663–667.

Marsalek, J., Chocat, B., 2001, International Report on Stormwater Management, World Water Congress, IWA, Berlin, October 15-19, 2001.

Martin, E., 1988, Effectiveness of an urban runoff detention pond/wetland system. *Journal of Environmental Engineering* 114(4):810-827.

Maryland Department of the Environment (MDE), 2000, *Maryland Stormwater Design Manual* http://www.mde.state.md.us/environment/wma/stormwater-manual. Accessed August 19, 2002.

Maxted, J, Shaver, E., 1997, The use of retention basins to mitigate stormwater impacts on aquatic life. In L.A. Roesner (Editor), *Effects of Watershed Development and Management of Aquatic Ecosystems,* American Society of Civil Engineers, New York, 494-512.

May, C., Horner, H., 2002, The Limitations of Mitigation-Based Stormwater Management in the Pacific Northwest and the Potential of a Conservation Strategy based on Low-Impact Development Principles. *Proceedings of the Ninth International Conference on Urban Drainage,* September 8-13, 2002.

May, C., Horner, R., Karr, J., Mar, B., Welch, E., 1997, Effects of urbanization on small streams in the Puget Sound lowland ecoregion, *Watershed Protection Techniques* 2(4): 483-494.

McLean, N., 2001, A study of the effectiveness of two SUDS ponds serving highway runoff. In Jefferies C. (Editor), *SUDS Monitoring Programme–Report and Interim Outcomes Scotland and Northern Ireland Forum for Environmental Research Report No. SR(00)* 10 August.

Mays, L.W., 2001, *Stormwater Collection Systems Design Handbook.* McGraw-Hill, Two Penn Plaza, New York.

Metropolitan Washington Council of Governments (MWCOG), 1993, *The Quality of Trapped Sediments and Pool Water Within Oil Grit Separators in Suburban Maryland.* Interim Report. Washington, D.C.

Mikkelsen, P., Viklander, M., Linde, J., Malmqvist, P., 2001, BMPs in Urban Stormwater Management in Denmark and Sweden, in proceedings of an Engineering Foundation Conference: Linking Stormwater BMP Designs and Performance to Receiving Water Impact Mitigation, August 19-24, Snowmass Village, CO, pp. 354-368.

Mikkelsen P.S., Hafliger M., Ochs M., Jacobsen P., Tjell J.C., Boller M., 1997, Pollution of soil and groundwater from infiltration of highly contaminated stormwater—a case study. *Water Science Technology,* 36(8-9): 325–330.

Minagawa, K., 1990, The stormwater infiltration system in housing complexes and the follow-up survey. In Y. Iwasa and T. Sueishi (Editors), *Drainage Systems and Runoff Reduction,* (Proc. 5th Int. Conf. on Urban Storm Drainage, Vol. 2), Osaka, Japan, 771-776.

Mineart, P., Singh, S., 1994, *Storm Inlet Pilot Study.* Alameda County Urban Runoff Clean Water Program, Oakland, CA.

Monster, N.J., Leeflang, M.J., 1996, *On the Design of Percolation Facilities: Communication of the Department of Sanitary Engineering and Water Management.* Delft University of Technology, Netherlands ISSN 0169-6246.

N.C. Department of Environment and Natural Resources, 1999, *Stormwater Best Management Practices.* Division of Water Quality Water Quality Section. Department of Environment and Natural Resources, NC.

Niemczynowicz, J., and Hogland, W., 1987, Test of porous pavements performed in Lund, Sweden. In B.C. Yen (Editor), *Topics in Drainage Hydraulics and Hydrology,* International Association for Hydraulic Research, Madrid, Spain (Secretariat, IAHR), 19–80.

Nightingale, H.I., 1989, Urban storm runoff effects on exchangeable cations and percolating water constituents below recharge basins. *Soil Science, V. 148, p. 39-45.*

Nightingale, H.I., 1987a, Water quality beneath urban runoff water management basins, *Water Resources Bulletin*, Vol. 23, p. 197-205.

Nightingale, H.I., 1987b, Accumulation of As, Ni, Cu, and Pb in retention and recharge basin soils from urban runoff, *Water Resources Bulletin*, Vol. 23, p. 663-672.

Nightingale, H.I., 1987c, Organic pollutants in soils of retention/recharge basins receiving urban runoff water, *Soil Science* Vol. 148, pp. 39-45.

Nightingale, H.I., 1975, Lead, zinc, and copper in soils of urban storm-runoff retention basins, *American Water Works Assoc. Journal*. Vol. 67, p. 443-446.

Nightingale, H.I., Harrison, D., Salo, J.E., 1985, an evaluation technique for ground-water quality beneath urban runoff retention and percolation basins, *Ground Water Monitoring Review*, Vol. 5, No. 1, pp. 43-50.

Northern Virginia Planning District Commission (NVPDC) and Engineers and Surveyors Institute, 1992, *Northern Virginia BMP Handbook*. Annandale, VA.

O'Shea, M., Borst, M., Nietch, C., 2002, The Role of Stormwater BMPs in Mitigating the Effects of Nutrient Overenrichment in the Urban Watershed. In *Global Solutions for Urban Drainage,* proceedings of the Ninth International Conference on Urban Drainage, September 8-13, 2002.

Oakland, P.H., 1983, An evaluation of stormwater pollutant removal through grassed swale treatment. In *Proceedings of the International Symposium of Urban Hydrology, Hydraulics and Sediment Control, Lexington, KY*. pp. 173-182.

Oberts, G.L., 1994, Performance of stormwater ponds and wetlands in winter. *Watershed Protection Techniques* 1(2):64-68.

Oberts, G.L., Wotzka, P.J., Hartsoe, J.A., 1989, *The Water Quality Performance of Select Urban Runoff Treatment Systems*. Publication No. 590-89-062a. Prepared for the Legislative Commission on Minnesota Resources, Metropolitan Council, St. Paul, MN.

Oberts, G.L., L. Wotzka, 1988, The water quality performance of a detention basin wetland treatment system in an urban area. In *Nonpoint Source Pollution: Economy, Policy, Management and Appropriate Technology*. American Water Resources Association, Middleburg, VA.

Occoquan Watershed Monitoring Laboratory, 1983, *Final Report: Metropolitan Washington Urban Runoff Project*. Prepared for the Metropolitan Washington Council of Governments, Washington, DC, by the Occoquan Watershed Monitoring Laboratory, Manassas, VA.

Ontario Ministry of the Environment, 1991, *Stormwater Quality Best Management Practices*. Marshall Macklin Monaghan Limited, Toronto, Ontario.

Osborne, K., Barrett, M., Maidment, D., 2000, *A Water Quality GIS Tool for the City of Austin Incorporating Non Point Sources and Best Management Practices,* Center for Research in Water Resources Online Report 00-01 (http://www.crwr.utexas.edu/reports/2000/rpt00-10.shtml), University of Texas at Austin.

Perry R., McIntyre, A.E., 1986, Impact of motorway runoff upon surface water quality. In *Proceedings of a Conference on the Effects of Land Use on Fresh Water*, Ellis Horwood, Chichester, U.K.

Phillips, I.R., 1998, Use of soil amendments to reduce nitrogen, phosphorus and heavy metal availability. *Journal of Soil Contamination* 7(2), 191-212.

Pitt, R., 1996, The Control of Toxicants at Critical Source Areas. Presented at the ASCE/Engineering Foundation Conference, Snowbird, UT, August 1996.

Pitt, R., Burton, G., 2001, Methods for the Assessment of Urban Wet-Weather Flow Impacts, in proceedings of an Engineering Foundation Conference: Linking Stormwater BMP Designs and Performance to Receiving Water Impact Mitigation, August 19-24, Snowmass Village, CO, 316-333.

Pitt, R., Lilburn, M., Nix, S., Durrans, S.R., Burian, S., Voorhees, J., Martinson, J., 2000, *Guidance Manual for Integrated Wet Weather Flow (WWF) Collection and Treatment Systems for Newly Urbanized Areas (New WWF Systems)*. U.S. EPA, Office of Research and Development, Cincinnati, OH.

Pitt, R., Lilburn, M., Burian, S., 1997, *Storm Drainage Design for the Future: Summary of Current U.S. EPA Research.* American Society of Civil Engineers Technical Conference, Gulf Shores, AL, July 1997.

Pitt, R., Clark, S., Parmer, K., 1994, *Potential Groundwater Contamination from Intentional and Nonintentional Stormwater Infiltration,* EPA/600/R-94/05 1, Risk Reduction Engineering Laboratory, U.S. EPA, Cincinnati, OH.

Pitt, R., McLean, J., 1986, *Toronto Area Watershed Management Strategy Study: Humber River Pilot Watershed Project.* Ontario Ministry of Environment, Toronto, ON.

Pitt, R., Bissonnette, P., 1984, *Bellevue Urban Runoff Program Summary Report*. U.S. Environmental Protection Agency, Water Planning Division, Washington, D.C.

Pratt, C.J., 2001, *Sustainable Urban Drainage. A Review of Published Material on the Performance of Various SUDS Devices.* Construction Industry Research and Information Association (CIRIA). Prepared for the UK Environment Agency. Coventry University, U.K. www.ciria.org.uk/suds/pdf/suds_review-jan02.pdf, accessed August 19, 2002.

Pratt C.J., 1996, Research and development in methods of soakaway design. *Journal CIWEM,* 10(1): 47-51.

Pratt, C.J., 1995, Infiltration drainage—case studies of U.K. practice. *Project Report 22*, Construction Industry Research and Information Association, London, SW1P 3AU; also known as *National Rivers Authority R & D Note 485.*

Price, M., 1994, Drainage from Roads and airfields to soakaways: groundwater pollutant or valuable recharge? *Journal of the Institution of Water and Environmental Management*, Vol. 8, 468-479.

Pratt, C.J., 1990, Permeable pavements for stormwater quality enhancement. In H.C. Torno (Editor), *Urban Stormwater Quality Enhancemen —Source Control, Retrofitting and Combined Sewer Technology.* ASCE, ISBN 087262 7594, 131–155.

Pratt C. J., Mantle J.D.G., Schofield, P.A., 1990, Porous pavements for flow and pollutant discharge control. In *Proceedings of the Fifth International Conference on Urban Storm Drainage,* Osaka, Japan, 839-844.

Pratt C. J., Newman A., Bond P.C., 1998, Mineral oil bio-degradation within a permeable pavement: long-term observations. In *Proceedings of the Third International Conference on Innovative Technologies in Urban Storm Drainage,* Lyon, France, Volume 1, 501-508.

Prince George's County Department of Environmental Resources (PGDER), 2001, *The Bioretention Manual.* Programs and Planning Division, Upper Marlboro, MD.

Prince George's County Department of Environmental Resources (PGDER), 1997, *Bioretention Monitoring: Preliminary Data Analysis.* Division of Environmental Management, Watershed Protection Branch. Landover, MD.

Prince Georges County Department of Environmental Resources (PGDER), 1993, *Design Manual for Use of Bioretention in Stormwater Management.* Division of Environmental Management, Watershed Protection Branch. Landover, MD.

Raimbault, G., 1997, French developments in reservoir structures and sustainable water resources in the 21st century. Malmo, Sweden.

Robertson, B., Pitt, R., Ayyoubi, A., Field, R., 1995, A Multi-Chambered Stormwater Treatment Train. In *Proceedings of the Engineering Foundation Conference: Stormwater NPDES-Related Monitoring Needs,* Mt. Crested Butte, Colorado, August 7-12, 1994, American Society of Civil Engineers, New York.

Roesner, L.A., Howard, R.M., Mack, B.W., Ramdett C.A., 1997, Integrating Stormwater management into Urban Planning in Orlando, Florida. Sustaining Urban Water Resources in the 21st Century, Engineering Foundation Conference, Malmo, Sweden. 8-12 September 1997.

Sansalone, J.J., Smithson, J.A., Koran, J.M., 1998, Development and testing of a partial exfiltration trench for in-situ treatment of highway drainage. *Transportation Research Record,* Vol 1647, pp 34-42.

Santana, F., Wood, J., Parsons, R., Chamberlain, S., 1994, *Control of Mosquito Breeding in Permitted Stormwater Systems.* Prepared for Southwest Florida Water Management District, Brooksville, FL.

Saunders, G., Gilroy, M., 1997, *Treatment of Nonpoint Source Pollution with Wetland/Aquatic Ecosystem Best Management Practices.* Texas Water Development Board, Lower Colorado River Authority, Austin, TX.

Schlüter, W., Jefferies, C., 2001, Monitoring the outflow from a porous car park *Proc. First National Conference on Sustainable Drainage Systems,* Coventry, UK.

Schlüter, W., Spitzer, A, Jefferies, C., 2002, Performance of three sustainable urban drainage systems in East Scotland. Paper for 9th International Conference on Urban Drainage, Portland, OR.

Schroeder, R.A., 1995, *Potential for Chemical Transport Beneath a Storm-Runoff Recharge (Retention) Basin for an Industrial Catchment in Fresno, CA*, USGS Water-Resource Investigations Report 93-4140.

Schueler, T., 1997a, Comparative pollutant removal capability of urban BMPs: A reanalysis. *Watershed Protection Techniques* 2(4):515-520.

Schueler, T., 1997b, Influence of groundwater on performance of stormwater ponds in Florida. *Watershed Protection Techniques* 2(4):525-528.

Schueler, T., 1994, Developments in sand filter technology to improve stormwater runoff quality. *Watershed Protection Techniques* 1(2):47-54.

Schueler, T.R., 1992, *A Current Assessment of Urban Best Management Practice*s. Metropolitan Washington Council of Governments. Southeastern Wisconsin Regional Planning Commission.

Schueler, T.R., 1987, *Controlling Urban Runoff: A Practical Manual for Planning and Designing Urban BMPs*. Metropolitan Washington Council of Governments, Washington, D.C.

Schueler, T.R., Kumble, P.A., Harity, M.A., 1992, *A Current Assessment of Urban Best Management Practices: Techniques for Reducing Non-Point Source Pollutants in the Coastal Zone.* Metropolitan Washington Council of Governments, Washington, D.C.

Schueler, T.R., Lugbill, J., 1990, *Performance of Current Sediment Control Measures at Maryland Construction Sites.* Metropolitan Washington Council of Governments. Pub. No. 89705. Department of Environmental Programs, Metropolitan Washington Council of Governments, Washington, D.C.

Seattle Metro and Washington Department of Ecology, 1992, *Biofiltration Swale Performance: Recommendations and Design Considerations*. Publication No. 657. Water Pollution Control Department, Seattle, WA.

Scottish Environment Protection Agency (SEPA), Environment Agency (EA), 1999; 2001, *Sustainable Urban Drainage Systems: An Introduction.* Planning & Development Services, Bournemouth, U.K. ISBN 1-901322-12-8.

Shutes, R., Revitt, D., Scholes, L., Forshaw, M., Winter, B., 2001, An experimental constructed wetland system for the treatment of highway runoff in the U.K. *Water Science Technology* 44: 11-12.

Southeastern Wisconsin Regional Planning Commission (SWRPC), 1991, *Costs of Urban Nonpoint Source Water Pollution Control Measures.* Technical report no. 31. Southeastern Wisconsin Regional Planning Commission, Waukesha, WI.

Stewart, W., 1992, *Compost Stormwater Treatment System.* W&H Pacific Consultants. Draft Report. Portland, OR. Also in Innovative Leaf Compost System Used to Filter Runoff at Small Sites in the Northwest. *Watershed Protection Techniques*. Center for Watershed Protection. February 1994. Vol. 1(1): 13-14.

Stormwater Management, 1997, Massachusetts Department of Environmental Protection 1997. *Stormwater Management Volume Two: Stormwater Technical Handbook*, Boston, MA.

Strecker, E., Urbonas, B., 2001, Assessing Receiving Water Effects of Urban Stormwater BMPs. In proceedings of an Engineering Foundation Conference: Linking Stormwater BMP Designs and Performance to Receiving Water Impact Mitigation, August 19-24, Snowmass Village, CO, 426-437.

Strecker, E., 1994, Constituents and methods for assessing BMPs. *Proc. Engrg. Found. Conf. on Storm Water Related Monitoring Needs*, ASCE, New York.

Stribling, J., Cummins, J., Galli, J., Meigs, S, Coffman, L., and Cheng, Mow-Soung, 2001, Relating Instream Biological Condition to BMPs in Watershed. In proceedings of an Engineering Foundation Conference: Linking Stormwater BMP Designs and Performance to Receiving Water Impact Mitigation, August 19-24, Snowmass Village, CO, 60-77.

SUDSWP, 2000, *Sustainable Urban Drainage Systems: Design Manual for England and Wales*. CIRIA, 2000.

SUDSWP, 1999, *Sustainable Urban Drainage Systems: Design Manual for Scotland and Northern Ireland*. CIRIA, 1999, Published for the Sustainable Urban Drainage Scottish Working Party.

Swietlik, W., 2001, Urban Aquatic Life Use—A Regulatory Perspective. In proceedings of an Engineering Foundation Conference: Linking Stormwater BMP Designs and Performance to Receiving Water Impact Mitigation, August 19-24, Snowmass Village, CO, pp. 60-77.

Todorovic, Z., Butler, D., Maksimovic, C., Prokrajac, D., and Deletic, A., 2002, Simulation of the long-term hydraulic behavior of stormwater infiltration trenches. Paper to be published in *Urban Water Journal*, Elsevier Science.

U.S. Environmental Protection Agency (U.S. EPA). 1999a, *Storm Water Technology Fact Sheet: Water Quality Inlets,* EPA 832-F-99-029, Office of Water, Washington, D.C.

U.S. Environmental Protection Agency (U.S. EPA), 1999b, *Storm Water Technology Fact Shee*t, *Vegetated Swales.* 832-F-99-006. Office of Water, Washington, D.C.

U.S. Environmental Protection Agency (U.S. EPA), 1999c, *National Menu of Best Management Practices for Stormwater Phase* [http://cfpub.epa.gov/npdes/stormwater/menuofbmps-/post_7.cfm] and [http://www.epa.gov/owm/mtb/infltrenc-.pdf] accessed August 22, 2002.

U.S. Environmental Protection Agency (U.S. EPA), 1999d, *Storm Water Technology Fact Sheet: Bioretention.* EPA 832-F-99-012. Office of Water, Washington, D.C.

U.S. Environmental Protection Agency (U.S. EPA), 1999e, *Storm Water Technology Fact Sheet: Wet Detention Ponds.* EPA 832-F-99-048. Office of Water, Washington, D.C.

U.S. Environmental Protection Agency (U.S. EPA), 1999f, *Storm Water Technology Fact Sheet: Sand Filters.* EPA 832-F-99-007. Office of Water, Washington , D.C.

U.S. Environmental Protection Agency (U.S. EPA), 1995, *Economic Benefits of Runoff Controls.* U.S. Environmental Protection Agency, Office of Wetlands, Oceans, and Watersheds, Washington, D.C.

U.S. Environmental Protection Agency (U.S. EPA), 1993, *Guidance Specifying Management Measures for Sources of Nonpoint Pollution in Coastal Waters,* Report 840-B-92-002, Office of Water, Washington, D.C.

U.S. Environmental Protection Agency (U.S. EPA), 1983, *Results of the Nationwide Urban Runoff Program: Volume—Final Report,* WH-554, Water Planning Division, Washington, D.C.

Urbonas, B., Carlson J., Vang, B., 1994, *Joint Pond-Wetland System in Colorado.* Denver Urban Drainage and Flood Control District, Denver, CO.

Urbonas, B.R., 1999, Design of a sand filter for stormwater quality enhancement, *Water Environment Research, 71*(1): 102-113.

Urbonas, B.R., 1994, Parameters to report with BMP monitoring data. *Proc. Engrg. Found. Conf. on Storm Water Monitoring-Related Monitoring Needs*, American Society of Civil Engineers (ASCE), New York.

Wang, T., Spyridakis, D., Mar, B., Horner, R., 1981, *Transport, Deposition and Control of Heavy Metals in Highway Runoff.* FHWA-WA-RD-39-10. University of Washington, Department of Civil Engineering, Seattle, WA.

Wanielista, M, Gennaro, R., Bell, J., Johnson, W., 1978, *Shallow Roadside Ditches for Stormwater Purification*, Florida Department of Transportation Report. FL-ER-3-78. Tallahassee, FL.

Warnaars, E., Larsen, A.V., Jacobsen, P., Mikkelsen, P.S., 1999, Hydrologic behaviour of stormwater infiltration trenches in a central urban area during 23/4 years of operation. *Water Science Technology*, 39(2), 217-224.

Washington State Department of Ecology (DOE), 1992, *Stormwater Management Manual for the Puget Sound Basin*, Washington State Department of Ecology, Olympia, WA.

Washington State Department of Transportation (WSDOT), 1995, *Highway Runoff Manual*, Washington State Department of Transportation, Olympia, WA.

Watershed Management Institute (WMI), 1997, *Operation, Maintenance, and Management of Stormwater Management Systems*. Prepared for U.S. Environmental Protection Agency, Office of Water, Washington, D.C., by the Watershed Management Institute, Ingleside, MD.

Watershed Protection Techniques, 1996, Ditches or Biological Filters? Classifying the Pollutant Removal Performance of Open Channels. Technical Notes-Urban Best Management Practices. *Watershed Protection Techniques* 2(2) Spring 1996.

Welborn, C., Veenhuis, J., 1987, *Effects of Runoff Controls on the Quantity and Quality of Urban Runoff in Two Locations in Austin, TX.* USGS Water Resources Investigations Report No. 87-4004. U.S. Geological Survey, Reston, VA.

Wigington, P.J., Jr., Randall, C.W., Grizzard, T.J., 1983, Accumulation of selected trace metals in soils of urban runoff detention basins, *Water Resources Bulletin*, Vol. 19, No. 5, pp. 709-718.

Wild, T.C., Jefferies, C., D'Arcy, B.J., 2002, *SUDS in Scotland—The Scottish SUDS Database, Report SR(02)09,* Scotland and Northern Ireland Forum for Environmental Research, Edinburgh.

Wu, J., 1989, *Evaluation of Detention Basin Performance in the Piedmont Region of North Carolina.* Report No. 89-248. North Carolina Water Resources Research Institute, Raleigh, NC.

Young, G.K., Stein, S., Cole, P., Kammer, T., Graziano, F., Bank, F, 1996, *Evaluation and Management of Highway Runoff Water Quality*, Publication No. FHWA-PD-96-032, U.S. Department of Transportation, Federal Highway Administration, Office of Environment and Planning. Washington, D.C.

Yousef, Y., Hitved-Jacobsen, T., Wanielista, M., Harper, H., 1987, Removal of contaminants in highway runoff flowing through swales, *Science of the Total Environment*, 59, 391-399.

Yousef, Y., Wanielista, M., Harper, H., 1986, Design and Effectiveness of Urban Retention Basins. In B. Urbonas and L.A. Roesner (Editors), *Urban Runoff Quality—Impact and Quality Enhancement Technology.* American Society of Civil Engineers (ASCE), New York, 338-350.

Yousef, Y., Wanielista, M., Harper, H., 1985, Removal of highway contaminants by roadside swales. *Transportation Research Record*, 1017: 62-68.

Yousef, Y., Wanielista, M., Harper, H., Pearce, D., Tolbert, R., 1985, *Best Management Practices: Removal of Highway Contaminants by Roadside Swales.* University of Central Florida and Florida Department of Transportation, Orlando, FL.

Yu, S., Barnes, S., Gerde, V., 1993, *Testing of Best Management Practices for Controlling Highway Runoff.* FHWA/VA-93-R16. Virginia Transportation Research Council, Charlottesville, VA.

WASTEWATER UTILITY

Alabama
Montgomery Water Works & Sanitary Sewer Board

Alaska
Anchorage Water & Wastewater Utility

Arizona
Gila Resources

Glendale, City of, Utilities Department

Mesa, City of

Peoria, City of

Phoenix Water Services Department

Pima County Wastewater Management

Arkansas
Little Rock Wastewater Utility

California
Calaveras County Water District

Central Contra Costa Sanitary District

Contra Costa Water District

Crestline Sanitation District

Delta Diablo Sanitation District

Dublin San Ramon Services District

East Bay Dischargers Authority

East Bay Municipal Utility District

El Dorado Irrigation District

Fairfield-Suisun Sewer District

Fresno Department of Public Utilities

Irvine Ranch Water District

Las Virgenes Municipal Water District

Livermore, City of

Lodi, City of

Los Angeles, City of

Los Angeles County, Sanitation Districts of

Napa Sanitation District

Orange County Sanitation District

Palo Alto, City of

Riverside, City of

Sacramento Regional County Sanitation District

San Diego Metropolitan Wastewater Department, City of

San Francisco, City & County of

San Jose, City of

Santa Barbara, City of

Santa Rosa, City of

South Bayside System Authority

South Coast Water District

South Orange County Wastewater Authority

Union Sanitary District

West Valley Sanitation District

Colorado
Boulder, City of

Colorado Springs, Utilities

Littleton/Englewood Water Pollution Control Plant

Metro Wastewater Reclamation District, Denver

Connecticut
The Mattabasset District

New Haven, City of, WPCA

District of Columbia
District of Columbia Water & Sewer Authority

Florida
Broward, County of

Fort Lauderdale, City of

Gainesville Regional Utilities

JEA

Miami-Dade Water & Sewer Authority

Orange County Utilities Department

Orlando, City of

Reedy Creek Improvement District

Seminole County Environmental Services

St. Petersburg, City of

Stuart Public Utilities

Tallahassee, City of

Tampa, City of

Toho Water Authority

West Palm Beach, City of

Georgia
Atlanta Department of Watershed Management

Augusta, City of

Clayton County Water Authority

Cobb County Water System

Columbus Water Works

Fulton County

Gwinnett County Department of Public Utilities

Macon Water Authority

Savannah, City of

Hawaii
Honolulu, City and County of

Illinois
American Bottoms Wastewater Treatment Plant

Greater Peoria Sanitary District

Kankakee River Metropolitan Agency

Metropolitan Water Reclamation District of Greater Chicago

North Shore Sanitary District

Wheaton Sanitary District

Indiana
Fort Wayne, City of

Iowa
Ames, City of

Cedar Rapids Wastewater Facility

Des Moines Metro Wastewater Reclamation Authority

Iowa City, City of

Kansas
Johnson County Unified Wastewater Districts

Unified Government of Wyandotte County/Kansas City, City of

Kentucky
Louisville & Jefferson County Metropolitan Sewer District

Louisiana
Sewerage & Water Board of New Orleans

Maine
Bangor, City of

Portland Water District

Maryland
Anne Arundel County Bureau of Utility Operations

Howard County Department of Public Works

Washington Suburban Sanitary Commission

Massachusetts
Boston Water & Sewer Commission

Upper Blackstone Water Pollution Abatement District

Michigan
Ann Arbor, City of

Detroit, City of

Holland Board of Public Works

Lansing, City of

Owosso Mid-Shiawassee County WWTP

Saginaw, City of

Wayne County Department of Environment

Wyoming, City of

Minnesota
Rochester, City of

Western Lake Superior Sanitary District

Missouri
Independence, City of

Kansas City Missouri Water Services Department

Little Blue Valley Sewer District

Nebraska
Lincoln Wastewater System

Nevada
Henderson, City of

New Jersey
Passaic Valley Sewerage Commissioners

New York
New York City Department of Environmental Protection

Rockland County Solid Waste Management Authority/ Sewer District

North Carolina
Cary, Town of

Charlotte/Mecklenburg Utilities

Durham, City of

Metropolitan Sewerage District of Buncombe County

Orange Water & Sewer Authority

Ohio
Akron, City of

Butler County Department of Environmental Services

Columbus, City of

Metropolitan Sewer District of Greater Cincinnati

Northeast Ohio Regional Sewer District

Summit, County of

Oklahoma
Tulsa, City of

Oregon
Clean Water Services

Eugene/Springfield Water Pollution Control

Water Environment Services

Pennsylvania
Philadelphia, City of

University Area Joint Authority, State College

South Carolina
Charleston Commissioners of Public Works

Mount Pleasant Waterworks & Sewer Commission

Spartanburg Sanitary Sewer District

Tennessee
Cleveland, City of

Knoxville Utilities Board

Murfreesboro Water & Sewer Department

Nashville Metro Water Services

Texas
Austin, City of

Dallas Water Utilities

Denton, City of

El Paso Water Utilities

Fort Worth, City of

Gulf Coast Waste Disposal Authority

WERF SUBSCRIBERS

Houston, City of

San Antonio Water System

Trinity River Authority

Utah

Salt Lake City Corporation

Virginia

Alexandria Sanitation Authority

Arlington, County of

Fairfax County Virginia

Hampton Roads Sanitation District

Henrico, County of

Hopewell Regional Wastewater Treatment Facility

Loudoun County Sanitation Authority

Lynchburg Regional WWTP

Prince William County Service Authority

Richmond, City of

Rivanna Water & Sewer Authority

Washington

Edmonds, City of

Everett, City of

King County Department of Natural Resources

Seattle Public Utilities

Sunnyside, Port of

Yakima, City of

Wisconsin

Green Bay Metro Sewerage District

Madison Metropolitan Sewerage District

Milwaukee Metropolitan Sewerage District

Racine, City of

Sheboygan Regional Wastewater Treatment

Wausau Water Works

Australia

South Australian Water Corp.

Sydney Water Corp.

Water Corp. of Western Australia

Canada

Greater Vancouver Regional District

Toronto, City of, Ontario

Winnipeg, City of, Manitoba

Mexico

Servicios de Agua y Drenaje de Monterrey, I.P.D.

New Zealand

Watercare Services Limited

Singapore

Singapore Public Utilities Board

United Kingdom

Yorkshire Water Services Limited

STORMWATER UTILITY

California

Los Angeles, City of, Department of Public Works

Monterey, City of

Santa Rosa, City of

Colorado

Boulder, City of

Georgia

Griffin, City of

Iowa

Cedar Rapids Wastewater Facility

Des Moines Metro Wastewater Reclamation Authority

Kansas

Overland Park, City of

Kentucky

Louisville & Jefferson County Metropolitan Sewer District

Maine

Portland Water District

Minnesota

Western Lake Superior Sanitary District

North Carolina

Cary, Town of

Charlotte, City of, Stormwater Services

Pennsylvania

Philadelphia, City of

Tennessee

Chattanooga Storm Water Management

Washington

Seattle Public Utilities

STATE

Arkansas Department of Environmental Quality

Fresno Metropolitan Flood Control District, Calif.

Urban Drainage & Flood Control District, Colo.

CORPORATE

ADS Environmental Services

The ADVENT Group Inc.

Alan Plummer & Associates

Alden Research Laboratory

Alpine Technology Inc.

Aqua-Aerobic Systems, Inc.

Aquateam–Norwegian Water Technology Centre A/S

BaySaver Inc.

BioVir Laboratories, Inc.

Black & Veatch

Boyle Engineering Corporation

BPR CSO

Brown & Caldwell

Burns & McDonnell

CABE Associates Inc.

The Cadmus Group

Camp Dresser & McKee Inc.

Carollo Engineers Inc.

Carpenter Environmental Associates Inc.

CDS Technologies Inc.

Chemtrac Systems Inc.

CH2M HILL

Damon S. Williams Associates, LLC

David L. Sheridan, P.C.

Earth Tech Inc.

Eco-Matrix

Environmental Engineers International

EMA Inc.

The Eshelman Company Inc.

Finkbeiner, Pettis, & Strout Inc.

Freese & Nichols, Inc.

Frontier Geosciences Inc.

ftn Associates Inc.

Fuss & O'Neill, Inc.

Gannett Fleming Inc.

Golder Associates Inc.

Greeley and Hansen LLC

Hazen & Sawyer, P.C.

HDR Engineering Inc.

HNTB Corporation

HydroQual Inc.

Ingersoll-Rand Energy Systems

Insituform Technologies Inc.

Institute for Environmental Technology & Industry, Korea

Jacobson Helgoth Consultants Inc.

Jason Consultants Inc.

Jordan, Jones, & Goulding Inc.

KCI Technologies Inc.

Kelly & Weaver, P.C.

Kennedy/Jenks Consultants

Komline Sanderson Engineering Corporation

Lawler, Matusky & Skelly Engineers, LLP

Limno-Tech Inc.

Lombardo Associates Inc.

Malcolm Pirnie Inc.

McKim & Creed

MEC Analytical Systems Inc.

Metcalf & Eddy Inc.

MWH

New England Organics

Odor & Corrosion Technology Consultants Inc. (OCTC)

ONDEO Degremont Inc.

Oswald Green, LLC

PA Government Services Inc.

Parametrix Inc.

Parsons

Post, Buckley, Schuh & Jernigan

R&D Engineering/Conestoga Rover & Associates

The RETEC Group

RMC, Inc. (Raines, Melton & Carella)

R.M. Towill Corporation

Ross & Associates Ltd.

Royce Technologies

SAIC Maritime Technical Group

Sear-Brown

Short Elliott Hendrickson, Inc.

Stormwater Management, Inc.

Synagro Technologies, Inc.

Tetra Tech Inc.

Trojan Technologies Inc.

URS Corporation

Veolia Water NATC

Wade-Trim Inc.

Weston Solutions Inc.

Woodard & Curran

WRc/D&B, LLC

WWETCO, LLC

Zoeller Pump Company

INDUSTRY

American Electric Power

ChevronTexaco Energy Research & Technology Company

The Coca-Cola Company

Dow Chemical Company

DuPont Company

Eastman Kodak Company

Merck & Company Inc.

ONDEO Services

Procter & Gamble Company

PSEG Services Corp.

Severn Trent Services Inc.

Shell Global Solutions

Thames Water Plc

United Water Services LLC

Note: List as of 1/31/04

WERF Product Order Form

As a benefit of joining the Water Environment Research Foundation, subscribers are entitled to receive one complimentary copy of all final reports and other products.Additional copies are available at cost (usually $10).To order your complimentary copy of a report, please write "free" in the unit price column.WERF keeps track of all orders.If the charge differs from what is shown here, we will call to confirm the total before processing.

Name ______________________ Title ______________________

Organization ______________________

Address ______________________

City ____________ State ____________ Zip Code ____________ Country ____________

Phone ____________ Fax ____________ Email ____________

Stock #	Product	Quantity	Unit Price	Total
			Postage & Handling	
			VA Residents Add 4.5% Sales Tax	
			Canadian Residents Add 7% GST	
			TOTAL	

Method of Payment:

(All orders must be prepaid.)

❑ Check or Money Order Enclosed

❑ Visa ❑ Mastercard ❑ American Express

Account No. ______________________ Exp. Date ____________

Signature ______________________

Shipping & Handling:

Amount of Order	United States	Canada & Mexico	All Others
Up to but not more than:	Add:	Add:	Add:
$20.00	$5.00*	$8.00	50% of amount
30.00	5.50	8.00	40% of amount
40.00	6.00	8.00	
50.00	6.50	14.00	
60.00	7.00	14.00	
80.00	8.00	14.00	
100.00	10.00	21.00	
150.00	12.50	28.00	
200.00	15.00	35.00	
More than $200.00	Add 20% of order	Add 20% of order	

*minimum amount for all orders

Note: Please make checks payable to the Water Environment Research Foundation.

To Order (Subscribers Only):

Log on to www.werf.org and click on the "Product Catalog."

Phone: (703) 684-2470
Fax: (703) 299-0742.

WERF
Attn: Subscriber Services
635 Slaters Lane
Alexandria, VA 22314-1177

To Order (Non-Subscribers):

Non-subscribers may be able to order WERF publications either through WEF (www.wef.org) or IWAP (www.iwapublishing.com).Visit WERF's website at www.werf.org for details.